SpringerBriefs in Computer Science

Series Editors

Stan Zdonik
Peng Ning
Shashi Shekhar
Jonathan Katz
Xindong Wu
Lakhmi C. Jain
David Padua
Xuemin Shen
Borko Furht
V. S. Subrahmanian
Martial Hebert
Katsushi Ikeuchi
Bruno Siciliano

For further volumes:
http://www.springer.com/series/10028

Daniel Schall

Service-Oriented Crowdsourcing

Architecture, Protocols and Algorithms

 Springer

Daniel Schall
Siemens Corporate Technology
Vienna
Austria

ISSN 2191-5768   ISSN 2191-5776   (electronic)
ISBN 978-1-4614-5955-2   ISBN 978-1-4614-5956-9   (eBook)
DOI 10.1007/978-1-4614-5956-9
Springer New York Heidelberg Dordrecht London

Library of Congress Control Number: 2012950384

Printed on acid-free paper

Springer is part of Springer Science+Business Media (www.springer.com)

Preface

Crowdsourcing has emerged as an important paradigm in human problem solving techniques on the Web. More often than noticed, programs outsource tasks to humans which are difficult to implement in software. Service-oriented crowdsourcing enhances these outsourcing techniques by applying the principles of service-oriented architecture (SOA) to the discovery, composition, and selection of a scalable human workforce. This book provides both an analysis of contemporary crowdsourcing systems such as Amazon Mechanical Turk and a statistical description of task-based marketplaces. In the following, a novel mixed service-oriented computing paradigm is introduced by providing an architectural description of the Human-Provided Services (HPS) framework and the application of social principles to human coordination and delegation actions. Then, the previously investigated concepts are extended to business process management integration including the extension of XML-based industry standards such as WS-HumanTask and BPEL4People and the instantiation of flexible processes in crowdsourcing environments.

Vienna, August 2012   Daniel Schall

Acknowledgments

The work presented in this book provides a consolidated description of the author's research in the field of human computation and crowdsourcing techniques. He started investigating crowdsourcing techniques in 2005 at Siemens Corporate Research in Princeton, NJ, USA. In 2006, he started his doctoral studies at the Vienna University of Technology (TU Wien) where he was employed as a project manager and research assistant. At that time he was involved in the EU FP6 project inContext (interaction and context-based technologies for collaborative teams) and defined a number of key principles such as the notion of Human-Provided Services and algorithms for context-sensitive expertise mining. In the following, the author worked as a Senior Research Scientist also at TU Wien where he was the principle investigator of efforts related to crowdsourcing techniques and mixed service-oriented systems. During this time period he was involved in a number of projects including the EU FP7 projects collaboration and interoperability for networked enterprises (COIN) and compliance-driven models, languages, and architectures for services (COMPAS). During his time at TU Wien, he published more than 50 scientific publications in highly ranked journals and renown magazines including the IEEE Transaction on Services Computing, IEEE Computer, IEEE Internet Computing, Data and Knowledge Engineering, Distributed and Parallel Databases, Social Network Analysis and Mining, Information Systems, as well as numerous world class conferences including the International Conference on Business Process Management, the International Conference on Services Computing, the International Conference on Advanced Information Systems Engineering, the International Conference on Social Informatics, the International Conference on Self-Adaptive and Self-Organizing Systems, or the International Conference on Engineering of Complex Computer Systems. The finalization of this book was carried out while the author has already been with Siemens Corporate Technology—a research division of the Siemens AG.

Acknowledgements

Contents

Acronyms

AMT   Amazon Mechanical Turk
API   Application Programming Interface
BPEL   Business Process Execution Language
B4P   Business Process Execution Language 4 People
BPM   Business Process Management
CFL   Crowd Flow
HIT   Human Intelligent Task
HPS   Human Provided Service
NFP   Nonfunctional Property
PFL   Process Flow
RFS   Request For Support
SBS   Software-Based Service
SOA   Service-Oriented Architecture
WSDL Web Services Description Language
WSHT Web Services Human Task
XML   Extended Markup Language

Chapter 1
Introduction

Abstract This chapter gives an introduction to human computation and crowdsourcing techniques. Next, the key features of human task marketplaces such as Amazon Mechanical Turk are briefly outlined. In the following, service-oriented crowdsourcing is motivated by giving an example. Finally, adaptive processes in the context of crowdsourcing are discussed and an outline of the book is given.

1.1 Overview

The shift toward the Web 2.0 allows people to write blogs about their activities, share knowledge in forums, write Wiki pages, and utilize social platforms to stay in touch with other people. Task-based platforms for human computation and crowdsourcing, including CrowdFlower [7], Google's Smartsheet [17], or Yahoo's Predictalot [11] enable access to the manpower of thousands of people on demand by creating human-tasks that are processed by the crowd. Human-tasks include activities such as designing, creating, and testing products, voting for best results, or organizing information. The notion of crowdsourcing describes an online, distributed problem solving and production model with increasingly interested business parties in the last couple of years [6]. Crowdsourcing follows the open world assumption [9] wherein peers interact and collaborate without being organized on a managerial/hierarchical model [5]. Thousands of individuals make their individual contributions to a body of knowledge and produce the core of our information and knowledge environment. One of the main motivations to outsource activities to a crowd is the potentially considerable spectrum of returned solutions. Furthermore, competition within the crowd ensures a certain level of quality.

According to [18], there are two dimensions in existing crowdsourcing platforms. The first categorizes the function of the platform. Currently these can be divided in communities (i) specialized on novel designs and innovative ideas, (ii) dealing with code development and testing, (iii) supporting marketing and sales strategies, and

D. Schall, *Service-Oriented Crowdsourcing*, SpringerBriefs in Computer Science, DOI: 10.1007/978-1-4614-5956-9_1, © The Author(s) 2012

(iv) providing knowledge support. Another dimension describes the crowdsourcing mode. Community brokers assemble a crowd according to the offered knowledge and abilities that bid for activities. Purely competition based crowdsourcing platforms operate without brokers in between. Depending on the platform, *incentives* for participation in the crowd are either monetary or simple credit-oriented. Even if crowdsourcing seems convenient and attracts enterprises with a scalable workforce and multilateral expertise, the challenges of crowdsourcing are a direct implication of human's ad-hoc, unpredictable behavior and a variety of interaction patterns.

1.2 Task Marketplaces

Task-based crowdsourcing platforms such as Amazon Mechanical Turk [2] (AMT) enable businesses to access the manpower of thousands of people on demand by posting human-task requests on Amazon's Web site. To date, AMT provides access to the largest group of workers available for processing Human Intelligent Tasks (HIT). Crowdsourcing platforms like AMT typically offer a user portal to manage HITs. Such tasks are made available via a *marketplace* and can be claimed by workers. In addition, most platforms offer application programming interfaces (APIs) to automate the management of tasks. However, from the platform point of view, there is currently very limited support in helping workers to identify relevant groups of tasks matching their interests. Also, as the number of both requesters issuing tasks and workers grows it becomes essential to define metrics assisting in the discovery of recommendable requesters. Some requesters may spam the platform by posting unusable tasks. A study from 2010 showed that 40 % of the HITs from new requesters are spam [10].

1.3 SOA for Crowdsourcing

Service-oriented architecture (SOA) is an emerging paradigm to realize extensible large-scale systems. As interactions and compositions spanning multiple enterprises become increasingly commonplace, organizational boundaries appear to be diminishing in future service-oriented systems. In such open and flexible enterprise environments, people contribute their capabilities in a service-oriented manner. We consider mixed service-oriented systems [12, 13] based on two elementary building blocks: (i) Software-Based Services (SBS), which are fully automated services and (ii) Human-Provided Services (HPS) [14] for interfacing with people in a flexible service-oriented manner. Here we discuss service-oriented environments wherein services can be added at any point in time. By following the open world assumption, humans actively shape the availability of HPSs by creating services. Interactions between HPSs are performed by using Web service-based technology (XML-based SOAP messages).

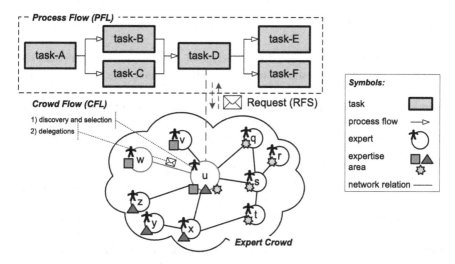

Fig. 1.1 Utilizing crowdsourcing in process flows

A motivating scenario for discovering members of the crowd in process-centric flows is depicted in Fig. 1.1.

The **Process Flow** (PFL) may be composed of single tasks that are either processed by corresponding Web services or are assigned to responsible persons. In this scenario, a task (task-D) may be outsourced to the crowd. This is done by preparing a *request for support* (RFS) containing various artifacts to be processed by the crowd and additional metadata such as time constraints and complexity of the task. The first step in a mixed service-oriented systems is to discover and select a suitable HPS. Discovery and selection is based on both, matching of functional capabilities (the service interface) and non-functional characteristics such as the degree of human expertise. In the depicted case, the actor u has been selected as the responsible HPS for processing the given request. The selection is based on u's expertise (visualized by the size of the node in the network), which is influenced by u's gradually evolving expertise and dynamically changing interests. The novelty of the approach is that members of the crowd may also interact with each other by, for example, simply delegating requests to other members (e.g., member u delegates the request to the peer w) or by splitting the request into sub-tasks that are assigned to multiple neighboring peers in the network. In our approach, the discovery of neighbors is based on the social structure of networks (e.g., friend or buddy lists). How decisions within the crowd are made (delegation or split of tasks) emerges over time due to changing interaction preferences and evolving capabilities of people (depicted as *expertise areas*). These dynamic interactions are defined as **Crowd Flow** (CFL). Flexible interaction models allow for the natural evolution of communities based on skills and interest. Our presented expertise mining approach and techniques help to address flexible interactions in crowdsourcing scenarios.

1.4 Adaptive Processes

Web services have paved the way for a new type of distributed system. Services let developers and engineers design systems in a modular manner, adhering to standard-ized interfaces. Services already play an important role in fulfilling organizations' business objectives because process stakeholders can design, implement, compose, and execute business processes using Web services as well as languages such as the Business Process Execution Language [4] (BPEL).

However, the BPEL specification was lacking a concept of (process) activities that are performed by human actors. Specifically the case that certain services in a process need to be provided by people is not covered. Recently, major software vendors have been working on standards addressing the lack of human interaction support in service-oriented systems. WS-HumanTask [3] (WS-HT) and BPEL4People [1] (B4P) were released to address the emergent need for human interactions in business-oriented processes. These standards specify languages for modeling human inter-actions, the lifecycle of humans tasks, and generic role models. Meanwhile, the Web-based *crowdsourcing* model called attempts to harnesses the creative solutions of a distributed network of individuals established with the goal to *outsource* tasks to workers [6, 9, 18]. This network of humans is typically an open Internet-based platform that follows the open world assumption and tries to attract members with different knowledge and interests. Large IT companies such as Amazon, Google, or Yahoo! have recognized the opportunities behind such mass collaboration sys-tems [8] for both improving their own services and as business case. While WS-HT and B4P have been defined to model human interactions in BPEL-based processes, it remains an open issue how to apply them to crowdsourcing. The WS-HT and B4P specifications need to be extended with Non-Functional Properties (NFPs) to ensure quality-aware crowdsourcing of human tasks.

1.5 Outline

This book is organized as follows. Both a statistical analysis of the Amazon Mechan-ical Turk marketplace and social network mining techniques of crowdsourcing task markets are presented in Chap. 2. In Chap. 3 Human-Provided Services (HPS) and mixed service-oriented systems are introduced. The integration of HPS and crowd-sourcing techniques into business process management are presented in Chap. 4. The book is concluded in Chap. 5.

The work presented in this book is based on the author's research performed over the last six years. The content is mainly based on the following journal publications:

- *Social Network Mining of Requester Communities in Crowdsourcing Markets* by D. Schall and F. Skopik (see [16]).

- *A Human-centric Runtime Framework for Mixed Service-oriented Systems* by D. Schall (see [13]).
- *Crowdsourcing Tasks to Social Networks in BPEL4People* by D. Schall, B. Satzger, and H. Psaier (see [15]).

References

1. Agrawal, A., et al.: WS-bpel extension for people (bpel4people), version 1.0 (2007)
2. Amazon mechanical turk. http://www.mturk.com/ (2012). Accessed 20 Aug 2012
3. Amend, M., et al.: Web services human task (WS-humantask), version 1.0 (2007)
4. Andrews, T., et al.: Business process execution language for web services, version 1.1 (2003)
5. Benkler, Y.: Coase's penguin, or linux and the nature of the firm. CoRR, cs.CY/0109077 (2001)
6. Brabham, D.: Crowdsourcing as a model for problem solving: an introduction and cases. Convergence **14**(1), 75 (2008)
7. Crowdflower. http://crowdflower.com/ (2012). Accessed 20 Aug 2012
8. Doan, A., Ramakrishnan, R., Halevy, A.Y.: Crowdsourcing systems on the World-Wide Web. Commun. ACM **54** (4), 86–96 (2011). doi:10.1145/1924421.1924442
9. Howe, J.: The rise of crowdsourcing. http://www.wired.com/wired/archive/14.06/crowds.html. June 2006
10. Ipeirotis, P.G.: Mechanical turk: now with 40.92% spam, (2010). http://bit.ly/mUGs1n. Accessed 20 Aug 2012
11. Predictalot. http://pulse.yahoo.com/y/apps/vU1ZXa5g/ (2012). Accessed 20 Aug 2012
12. Schall, D.: Human interactions in mixed systems—architecture, protocols, and algorithms. PhD thesis, Vienna University of Technology, Vienna (2009)
13. Schall, D.: A human-centric runtime framework for mixed service-oriented systems. Distrib. Parallel Databases **29**, 333–360 (2011). doi:10.1007/s10619-011-7081-z
14. Schall, D., Truong, H.-L., Dustdar, S.: Unifying human and software services in web-scale collaborations. IEEE Internet Comput. **12**(3), 62–68 (2008). doi:10.1109/MIC.2008.66
15. Schall, D., Satzger, B., Psaier, H.: Crowdsourcing tasks to social networks in BPEL4People. World Wide Web J. (2012). doi:10.1007/s11280-012-0180-6
16. Schall, D., Skopik, F.: Social network mining of requester communities in crowdsourcing markets. Soc. Netw. Anal. Min. (2012). doi:10.1007/s13278-012-0080-x
17. Smartsheet. http://www.smartsheet.com/ (2010). Accessed 20 Aug 2012
18. Vukovic, M.: Crowdsourcing for enterprises. In: Proceedings of the 2009 congress on services. IEEE Comput. Soc. 686–692 (2009)

Chapter 2
Crowdsourcing Task Marketplaces

Abstract In this chapter, we discuss detailed statistics of the popular Amazon Mechanical Turk (AMT) marketplace to provide insights in task properties and requester behavior. We present a model to automatically infer requester communities based on task keywords. Hierarchical clustering is used to identify relations between keywords associated with tasks. We present novel techniques to rank communities and requesters by using a graph-based algorithm. Furthermore, we introduce models and methods for the discovery of relevant crowdsourcing brokers who are able to act as intermediaries between requesters and platforms such as AMT.

Keywords Crowdsourcing · Mechanical turk · Hierarchical clustering · Community detection · Community ranking · Broker discovery

2.1 Introduction

In this chapter we define the notion of *communities* in the context of crowdsourcing. Communities are not predefined but emerge bottom-up based on posted tasks. Here we use keyword information applied to tasks to identify communities and community members (i.e., requesters). Hence, communities are mainly driven by *requesters*. For example, the keywords 'classification' and 'article' identify a community who makes tasks regarding the categorization of articles available. Managing the community standing of requesters in an automated manner helps to identify those requesters who contribute to a valuable marketplace.

In this chapter, we present the following key contributions:

- *Basic AMT Marketplace Statistics*. We thoroughly examine an AMT dataset and study properties regarding task distribution, rewarding, requester behavior and task keyword usage. The analysis of basic features and statistics provides the basis for the discovery of communities and the requester ranking model.

D. Schall, *Service-Oriented Crowdsourcing*, SpringerBriefs in Computer Science, DOI: 10.1007/978-1-4614-5956-9_2, © The Author(s) 2012

- *Keyword Clustering Approach.* Hierarchical clustering is used to identify relations between keywords associated with tasks, and finally requester communities demanding for workers in particular expertise areas. This is an important step toward a community ranking model. To our best knowledge, there is no existing work that shows how to automatically discover communities in task-based crowdsourcing marketplaces.
- *Community Ranking Model.* We propose link analysis techniques derived from popular Web mining algorithms to rank requesters and communities. This model helps to rate requesters with respect to their task involvement on AMT.
- *Broker Discovery Model.* We present a novel model for the discovery and ranking of *crowdsourcing brokers*. Brokers act as intermediaries between requesters and platform providers. The duty of brokers is to provide a specialized interface towards crowdsourcing platforms by the provisioning of additional services such as quality assurance or validation of task results.
- *Evaluation of the Community Ranking Model and Broker Discovery Approach.* Our evaluation and discussions are based on the properties of a real crowdsourcing marketplace.

This chapter is organized as follows. Section 2.2 outlines important related work. In Sect. 2.3 we highlight the basic properties of the AMT marketplace, including interactions and the system context model. This is the basis for Sect. 2.4, where we discuss a hierarchical clustering approach in order to group keywords and subsequently associate tasks. Using that model, we introduce a task requester and community ranking model. In Sect. 2.5 we present the broker discovery and ranking model. Section 2.6 details our experiments that are based on real data obtained from the AMT platform. Section 2.7 concludes the chapter.

2.2 Background

The notion of **crowdsourcing** was coined by Howe [27, 28] and is defined as *'the act of taking a job traditionally performed by a designated agent and outsourcing it to an undefined, generally large group of people in the form of an open call'.* The crowdsourcing paradigm [14, 43] has recently gained increased attention from both academia and industry, and is even considered for application in large-scale enterprises.

Crowdsourcing offers a attractive way to solve resource intensive tasks that cannot be processed by software [49]; typically all kinds of tasks dealing with matching, ranking, or aggregating data based on fuzzy criteria. Some concrete examples include relevance evaluation [1], evaluation of visual designs and their perception by large user groups [24], and ranking of search results [8]. Numerous further approaches deal with the seamless integration of crowds into business processes and information system architectures: CrowdDB [20] uses human input via crowdsourcing to process queries that neither database systems nor search engines can adequately answer.

Others study algorithms which incorporate human computation as function calls [35]. One of the largest and most popular crowdsourcing platforms is AMT. Besides tagging images and evaluating or rating objects, creating speech and language data with AMT [7] and the transcription of spoken language [36] are in the focus of the large application area of language processing and language studies [38]. Recently various platforms have been established that interface with and harness AMT, in order to provide more customized services, such as SmartSheet [56] and CrowdFlower [13].

Tagging is used to improve the navigation in folksonomies [21] and has been widely studied [22]. Applying tags to objects helps users to discover and distinguish relevant resources. For instance, users manually annotate their photos on Flickr [18] using tags, which describe the contents of the photo or provide additional contextual and semantical information. This feature is also utilized on the AMT platform, where tasks are described by tags, informing potential workers about the nature of a task and basic required skills. In contrast to predefined categories, tags allow people to navigate in large information spaces, unencumbered by a fixed navigational scheme or conceptual hierarchy. Previous works [54] investigated concepts to assist users in the tagging phase. Tags can also assist in creating relationship between semantic similarity of user profile entries and the social network topology [3].

Several approaches have been introduced, dealing with the construction of **hierarchical structures** of tags [15, 26, 55], generating user profiles based on collaborative tagging [37, 53], and collaborative filtering in general [25]. Our work aims at a similar goal by clustering tags and recommending categories of keywords to requesters and workers looking for interesting tasks. Our approach uses various methods and techniques from the information retrieval domain, including term-frequency metrics [46], measuring similarities [51], and hierarchical clustering [44].

With regards to **community and role detection**, community detection techniques can be used to identify trends in online social networks [9]. A context-sensitive approach to community detection is proposed in [5] whereas [45] proposes random walks to reveal community structure. Actors in large scale online communities typically occupy different roles within the social network [17]. The authors in [16] present methods for classification of different social network actors. Certain actors may act as moderators to separate high and low quality content in online conversations [34]. We specifically focus on the notion of *community brokers* who have the ability to assemble a crowd according to the offered knowledge [58]. Brokers in a sociological context may bridge segregated collaborative networks [52]. These community brokers could be ranked according to their betweenness centrality in social networks (see [33] for identifying high betweenness centrality nodes). The idea of structural holes, as introduced by Burt [6], is that gaps arise in online social networks between two individuals with complementary resources or information. When the two are connected through a third individual (e.g., the broker) the gap is filled, thereby creating important advantages for the broker. Competitive advantage is a matter of access to structural holes in relation to market transactions [6].

We position our work in the context of crowdsourcing with the focus on *requester communities*. Some works [29, 31] already studied the most important aspects of the

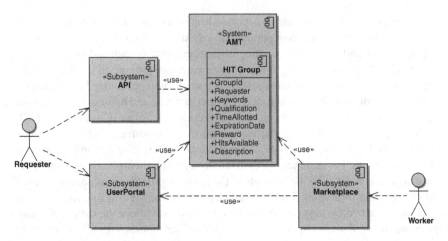

Fig. 2.1 Crowdsourcing system context model

AMT user community and describe analysis results of their structure and properties. While these works provide a basis for our experiments, we go important steps further:

1. We introduce *crowdsourcing communities* that are identified bottom-up through the analysis of the hierarchical keyword structure.
2. We develop a sophisticated link-based ranking approach to rank communities and requesters within the AMT community.
3. Here we propose the discovery of community brokers by adapting popular link mining techniques. The novelty of our approach is that the crowdsourcing broker discovery is based on query sensitive personalization techniques.

In the next section we introduce a basic task-based crowdsourcing model and discuss the statistics of the popular AMT crowdsourcing marketplace.

2.3 Basic Model and Statistics

2.3.1 System Context Overview

In this section we detail the basic system elements and user interactions. Figure 2.1 shows the high-level model and a set of generic building blocks. We illustrate the system context model by using the AMT platform and its HIT data model as an example of a task-based crowdsourcing marketplace.

- At the core, the **AMT** middleware offers the task management with a definition of the basic model for a **HIT Group**. The *GroupId* is a unique identifier of a HIT group. A HIT group encapsulates a number of HIT instances (*HitsAvailable*).

Workers can claim HIT instances within a group. The *Requester* identifier associates a task requester with a HIT group. Each HIT has a set of *Keywords* that will play a central role in subsequent discussions. The requester can define *Qualification* requirements such as geographical location. HITs are given a duration (specified as *TimeAllotted*) and an *ExpirationDate*. Workers receive a monetary *Reward* after successfully finishing a HIT instance. The *Description* attribute provides additional textual information.

- **Requesters** post tasks to the platform by using either the **User Portal** or a Web services based **API**. APIs help to automate the creation and monitoring of HITs. In addition, 3rd party crowdsourcing platform providers have the ability to build their own platforms on top of the AMT middleware.
- **Workers** are able to claim HIT instances from the **Marketplace** if they qualify for a given task. Additional constrains can be given by the requester, such as required skills or desired quality. Quality management is typically provided by 3rd party crowdsourcing platform providers (e.g., CrowdFlower) and not by the AMT system itself.

2.3.2 Marketplace Task Statistics

The techniques presented in this work are generally applicable to task-based crowdsourcing environments. To illustrate the application and rational behind our community discovery and ranking model, we will discuss the key features and statistics of real world crowdsourcing systems such as the AMT marketplace. We collected a HIT-dataset by periodically crawling AMT's Web site between February and August 2011 (in total seven months). The dataset contains 101027 HITs (5372355 HIT instances) and 5584 distinct requesters that were active during the time frame by making new HITs available.

Figure 2.2 shows the basic task statistics from the obtained dataset. In Fig. 2.2a we show the number of tasks and the number of requesters in a scatter plot with logarithmic scale on both axis (in short, log-log scale). The basic task-requester distribution follows the law that only few requesters post many tasks (the top-requester SpeechInk [57] posts 32175 HITs) while a large portion of requesters only post few tasks. A number of 2393 requesters (i.e. 43 %) only posts one task.

Next in Fig. 2.2b we show the number of tasks that require qualification versus tasks that do not require any particular qualification. The x-axis shows the number of task instances available within a HIT group and the y-axis depicts the number of tasks grouped by the amount of task instances available. Generally, more tasks require some sort of qualification like based on location ('Location is not in India') or based on qualification ('Image Transcription Description Qualification is greater than 88'). Thus, from the requester point of view, there is already some pre-selection of workers. However, AMT offers limited support to actually filter and rank tasks and requesters.

Figure 2.2c shows the time allotted to tasks (in minutes). The largest segment of tasks is concentrated around 60–100 min. This means that most tasks are relatively

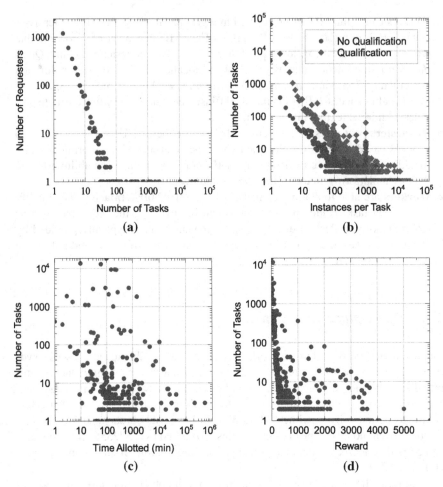

Fig. 2.2 Task statistics. **a** Number of tasks. **b** Qualification. **c** Time alloted. **d** Task reward

simple such as searching for email addresses or tagging of images. Finally, Fig. 2.2d shows the reward in cents (US currency) given for processing tasks (the x-axis is shown on a linear scale). The maximum reward given for a task is 60$. However, we find that most tasks have relatively little reward (26604 tasks have less than 55 cents reward).

Our community discovery and ranking approach uses task-keyword information as input for clustering of communities. Figure 2.3 shows the most important keyword statistics (all log-log scale). Figure 2.3a shows the number of keywords versus the number of requesters. The x-axis is based on the total number of keywords used by requesters. The distribution has its maximum (y-axis) at 4 keywords amounting for 758 requesters. Next, Fig. 2.3b depicts the average number of HIT keywords in relation to the number of requesters. By inspecting the maximum value (y-axis),

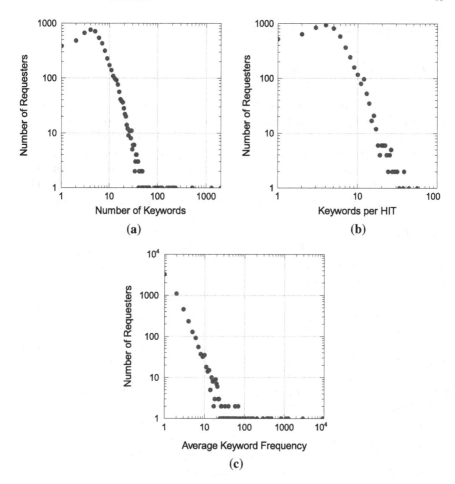

Fig. 2.3 Keyword statistics. **a** Keywords. **b** Keywords HIT. **c** Frequency

we observe that 935 requesters apply on average 4 keywords per HIT. The last keyword related statistic is shown in Fig. 2.3c depicting how often a particular keyword is used. By looking at the raw (unfiltered) keyword set, keywords with the highest frequency include company names of 3rd party platform providers such as SpeechInk, CastingWords or CrowdFlower.

However, a prerequisite for creating meaningful (hierarchical) clusters is a set of keywords that is not distorted by such keyword relations. Thus, we performed some filtering and cleaning of all stopwords ('and', 'on', 'is' and so forth) and also company names. Amongst those top-ranked keywords are 'survey', 'data', or 'collection'. Ranked by total reward, the keywords 'data' (79492$), 'transcribe' (55013$), 'search' (54744$), 'transcription' (47268$), 'collection' (43156$), and 'voicemail' (42580$) would be among the top ones.

2.4 Clustering and Community Detection

2.4.1 Clustering Approach

Current crowdsourcing platforms such as AMT offer limited search and navigation support for both requesters and workers. Workers are usually presented a long list of tasks to potentially work on and need to navigate from page-to-page to discover new and interesting tasks. Some workers may prefer to work on tasks regarding a specific topic of interest. Therefore, workers should be able to discover communities that post tasks regarding the desired topic. Requesters pick keywords they would like to apply to tasks freely and independently without following a particular convention or taxonomy. The positive aspect of a bottom-up approach (i.e., freely choosing keywords) is a domain vocabulary that may actually change based on the keywords chosen by the requesters. On the downside, problems include spelling mistakes, ambiguity, or synonyms because a large amount of different keywords may be used to describe the same type of task.

We propose *hierarchical clustering* to structure the flat set of task-based keywords into a hierarchy. The general idea is to first calculate the co-occurrence frequency of each keyword (how many times a particular keyword is used in combination with another keyword) and second group pairs of keywords into clusters based on a distance metric. Each HIT keyword starts in its own cluster. Subsequently pairs of clusters are merged by moving up the hierarchy. In other words, the correlation between keywords increases by moving from the top (root) to the bottom (leaves).

We have tested different distance metrics and configurations of the clustering algorithm. Based on our experiments, the following configuration yielded the best results (i.e., hierarchical structure). *Pairwise average-link clustering* merges in each iteration the pair of clusters with the highest cohesion. We used the *city-block distance*, alternatively known as the Manhattan distance, to measure the cohesiveness between pairs of clusters. In the conducted experiments, the input for the clustering algorithm was a set of about 300 keywords that have already been filtered as described in the previous section. Furthermore, we only used those keywords that had a co-occurrence frequency of at least 10 with some other keyword (minimum threshold). In total, the algorithm generates 328 clusters.

The next step in our approach is to create communities using the layout of the keyword-based hierarchy. This is shown in Algorithm 1. It is important to note that in Line 7 of the algorithm the keywords of all child-clusters are retrieved as well. To calculate the overlap in Line 9, the set intersection between KW_{HIT} and $KW_{Cluster}$ is divided by the set size $|KW_{Cluster}|$. Note that by associating collections of HITs to clusters (Line 16) we extend the notion of clusters to *communities* (i.e., an extended structure of a cluster with associated tasks and requesters). As a next step, we calculate basic statistics of the resulting community structure.

First, we show how many tasks requesters have in each cluster (Fig. 2.4). Since many requesters post only one task, also the count of requesters that have only one task in a cluster is high (757 requesters have one task in a cluster). In the middle

---

Algorithm 1 Creating communities using keyword hierarchy.

1: **input:** Set of Tasks, Keyword Hierarchy
2: **for each** HIT in Set of Tasks **do**
3:    **for each** Cluster in Hierarchy **do**
4:        // Keywords of HIT
5:        KW_{HIT} ← GetKeywords(HIT)
6:        // Keywords of Cluster and Children
7:        $KW_{Cluster}$ ← GetKeywords(Cluster)
8:        // Save Overlap
9:        Overlap(HIT, Cluster) ← $\frac{KW_{HIT} \cap KW_{Cluster}}{|KW_{Cluster}|}$
10:    **end for**
11:    // Sort Clusters by Overlap
12:    SortedList ← GetSortedClusterList(Overlap, HIT)
13:    // Pick highest ranked Cluster
14:    Cluster ← PickFirst(SortedList)
15:    // Add HIT to Collection associated with Cluster
16:    CollectionAdd(Cluster, HIT)
17: **end for**

---

segment, 744 requesters have 2 to 10 tasks in clusters. The high score is one requester with 2058 in one cluster.

A natural question when performing clustering is the quality of the resulting hierarchy. As mentioned before, we have evaluated the resulting clusters by looking at the groupings of keywords, which were consistent. Another possible metric for measuring the quality is the distance of similar tasks. Recall, each HIT is associated with a cluster (Algorithm 1, Line 16). Also, the hierarchy of clusters can be represented as a directed graph $G(V, E)$ where vertices V represent clusters and edges E the set of links between clusters (e.g., the root node points to its children and so forth). To understand whether the hierarchy represents a good structure, we sampled 1000 pairs of tasks randomly from the entire set of tasks and calculated the keyword-based similarity between the pair. Next, we calculated the Dijkstra shortest path distance between the pair of tasks using $G(V, E)$. The results are depicted by Fig. 2.5.

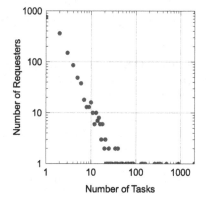

Fig. 2.4 Number of tasks in community clusters

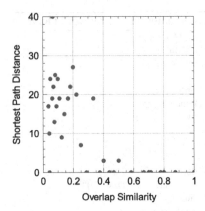

Fig. 2.5 Overlap versus distance in community structure

One can see that the hierarchy of clusters represents a good mapping between overlap similarity and shortest path distance: similar tasks have a low distance and tasks with little overlap similarity are very distant from each other in the graph G. Another positive effect of the proposed clustering and community discovery method (task association to clusters) is that spam or unclassified tasks can be identified.

2.4.2 Community-Based Ranking Model

In the previous section we investigated clustering methods to build communities using the keyword-based hierarchy. Here we attempt to answer the following question: which are the most important communities and who are the most recommendable requesters? The applications of the presented ranking approach are, for example, recommending relevant communities to both workers and requesters (e.g., to find interesting tasks) and also rating tasks of requesters with a high community standing. First, we need to detail the meaning of 'relevant communities' and 'community standing' of requesters. A relevant community is identified based on the authority of the requesters that post tasks to it. The community standing of requesters (i.e., authority) is established upon the relevancy of the communities the requester posts tasks to. Mathematically, the idea of this ranking model can be formalized using the notion of *hubs and authorities* as introduced in [32].

Formally, this recursive definition is written as

$$\mathscr{H}(c) = \sum_{(c,r)\in E_{CR}} w(r \to c)\mathscr{A}(r) \qquad (2.1)$$

$$\mathscr{A}(r) = \sum_{(c,r)\in E_{CR}} w(r \to c)\mathscr{H}(c) \qquad (2.2)$$

with $\mathscr{H}(c)$ being the hub score of the community $c \in V_C$ in the set of communities V_C, $\mathscr{A}(r)$ the authority score of the requester $r \in V_R$ in the set of requesters V_R, $(c, r) \in E_{CR}$ an edge in the community-requester bipartite graph $G_{CR}(V_C, V_R, E_{CR})$, and $w(r \to c)$ a weighting function based on the number of tasks posted by r in a community c. Notice, by posting tasks to a community c, an edge is established between the community c and the requester r.

2.5 Crowdsourcing Broker Discovery

There are multiple companies that provide marketplaces where users can post tasks that are processed by workers. Among the previously discussed AMT, there are also platform providers such as oDesk [39] and Samasource [47]. In contrast other companies, such as CrowdFlower [13] and ClickWorker [12] act as intermediaries allowing large businesses and corporations to not have to worry about framing and posting tasks to crowdsourcing marketplaces [41]. We call such intermediaries *brokers*. Brokers post tasks on behalf of other crowdsourcing requesters. Typically, such brokers offer additional services on top of platforms like AMT including quality control (e.g., see CrowdFlower [13]) or the management of Service-Level-Agreements (SLAs) [42].

In this work we provide a model for the discovery and ranking of brokers based on requester profile information. A requester's profile is created based on the task posting behavior and associated keywords. The requester profile contains a set of keywords and their frequency. The profile is defined as follows:

$$P_u = \left\{ \langle k_1, f_{k_1} \rangle, \langle k_2, f_{k_2} \rangle, \dots, \langle k_n, f_{k_n} \rangle \right\} \tag{2.3}$$

where $\langle k_n, f_{k_n} \rangle$ denotes the tuple of keyword k_n and its frequency f_{k_n}. Next we propose the creation of a directed *profile graph* $G_{PG}(V_C, E_{PG})$ that is created using the following algorithm:

Algorithm 2 Creating profile graph G_{PG} using requester profile information.

1: **input:** Set V_R of Requesters
2: **for each** Requester $u \in V_R$ **do**
3:    **for each** Requester $v \in V_R$ **do**
4:        **if** $u \neq v$ **then**
5:            // Calculate match between u and v
6:            $pm \leftarrow match(u, v)$
7:            **if** $(pm > \xi)$ or $(\xi = 1$ and $pm = \xi)$ **then**
8:                // Add profile relation to G_{PG}
9:                GraphAddRelation(u, v)
10:           **end if**
11:       **end if**
12:   **end for**
13: **end for**

The idea of our approach (as highlighted in Algorithm 2) is to establish a directed edge $(u, v) \in E_{PG}$ from u to v if there is high match (i.e., profile similarity) between u and v from u's point of view. The parameter ξ can be used to adjust the similarity threshold that the profile match pm must exceed to connect u and v through a profile relation edge.

$$match(u, v) = 1 - \sum_{k \in KW_u} w_k Max((f_k(u) - f_k(v))/f_k(u), 0) \qquad (2.4)$$

The calculation of the degree of match (see 2.4) is not symmetric. This is done because of the following reason. The requester v's profile P_v might exactly match u's profile P_u with $P_u \subseteq P_v$ ($match(u, v)$) but this may not be true when matching u's profile from v's point of view ($match(v, u)$). Suppose we calculate $match(u, v)$, a match of 1 means that v perfectly matches u's profile. Thus, it can be assumed that v has posted some tasks that are very similar to those posted by u. Therefore, it can be said that there is a high degree of interest similarity between u and v and v could potentially act as a broker for u. Again, keyword information associated with the requesters' HITs is used to create profiles through mining. Also keyword frequency is taken into account when calculating profile matches. A detailed description on the symbols depicted in (2.4) can be found in Table 2.1.

As mentioned before, a broker could submit the task on behalf of another requester and monitor the task's progress or could even segment the task into subtasks and submit the subtasks to one or more crowdsourcing platforms. At this stage, we focus on the discovery and ranking techniques of brokers without discussing the actual broker-requester interaction model. For instance, the management of SLAs in crowdsourcing environments has been addressed in our previous work [42] and is not the focus of this research.

Here we focus on the discovery and ranking of relevant brokers. Compared to the previously defined community requester graph G_{CR}, the profile graph G_{PG} consists only of a single type of nodes V_R. In this case, the requester importance is not influenced by the relevance of communities but rather by the degree of connectivity within the graph G_{PG}. A well-known and popular model to measure importance in directed networks is PageRank [40]. A advantage over the hubs and authority method [32] is that the PageRank model corresponds to a random walk on the graph.

Table 2.1 Description of profile matching calculation

Symbol	Description
$match(u, v)$	The matching of profiles between u and v. A value between [0, 1]
KW_u	The set of keywords used by u
f_k	The frequency of a keyword k. The frequency $f_k(u)$ is counted based on how many times the keyword k has been applied to tasks posted by u
w_k	The weight of a specific keyword k. The weight is calculated as $\frac{f_k}{\sum_{k_i} f_{k_i}}$

The PageRank $pr(u)$ of a node u is defined as follows:

$$pr(u) = \frac{\alpha}{|V_R|} + (1 - \alpha) \sum_{(v,u) \in E_{PG}} w(v, u) pr(v) \tag{2.5}$$

At a given node u, with probability α the random walk continues by following the neighbors $(v, u) \in E_{PG}$ connected to u and with probability $(1 - \alpha)$ the walk is restarted at a random node. The probability of 'teleporting' to any node in the graph is given as by the uniform distribution $\frac{1}{|V_R|}$. The weight of the edge (v, u) is given as $w(v, u)$. The default value for the transition probability between v and u is given as $\frac{1}{\texttt{outdegree}(v)}$. The function `outdegree` returns the count of the edges originating from v. The model can also be personalized by assigning non-uniform 'teleportation' vectors, which is shown in the following:

$$ppr(u; Q) = \alpha p(u; Q) + (1 - \alpha) \sum_{(v,u) \in E_{PG}} \frac{ppr(v)}{\texttt{outdegree}(v)} \tag{2.6}$$

The personalized PageRank $ppr(u; Q)$ is parameterized by the keyword based query Q. Instead of assigning uniform teleportation probabilities to each node (i.e., $\frac{1}{|V_R|}$), we assign preferences to nodes that are stored in $p(u; Q)$. This approach is similar to the topic-sensitive PageRank proposed by [23] (see also [10, 19, 30, 50]). Whereas in PageRank the importance of a node is implicitly computed relative to all nodes in the graph now importance is computed relative to the nodes specified in the personalization vector. The query Q is defined as a simple set of keywords $Q = \{k_1, k_2, \ldots, k_n\}$ that are selected to depict a particular topic(s) of interest. Algorithm 3 shows how to compute the values within the personalization vector $p(u; Q)$.

2.6 Experiments

The discussions on our evaluation and results in separated into two sections: first we discuss experiments of our community-based ranking model followed by discussions of the crowdsourcing broker ranking approach.

2.6.1 Community Discovery and Ranking

Here we discuss ranking results obtained by calculating $\mathcal{H}$ and $\mathcal{A}$ scores using the community-requester graph G_{CR}. Communities are visualized as triangular shapes (in blue color) and requesters are visualized as circles (in red color). The size of each shape is proportional to the $\mathcal{H}$ and $\mathcal{A}$ scores respectively. The line width of an edge is based on the weight $w(r \rightarrow c)$.

Algorithm 3 Assigning personalization vector $p(u; Q)$ based on query Q.

1: **input:** Set V_R of Requesters and query Q
2: // Variable $totalSum$ is used for normalization
3: $totalSum \leftarrow 0$
4: **for each** Requester $u \in V_R$ **do**
5:    $currentSum \leftarrow 0$
6:    **for each** Keyword $k^Q \in Q$ **do**
7:      **for each** Keyword $k \in KW_u$ **do**
8:        // If k^Q and k matches
9:        **if** $Equals(k^Q, k)$ and $f_k(u) > 0$ **then**
10:          // Add frequency $f_k(u)$ to $currentSum$
11:          $currentSum \leftarrow currenSum + f_k(u)$
12:        **end if**
13:      **end for**
14:    **end for**
15:    PersonalizationVectorAdd(u, $currentSum$)
16:    $totalSum \leftarrow totalSum + currentSum$
17: **end for**
18: // Normalize the values in $p(u; Q)$
19: **for each** Requester $u \in V_R$ **do**
20:    $weight \leftarrow$ PersonalizationVectorGet(u)
21:    PersonalizationVectorAdd(u, $weight/currentSum$)
22: **end for**

First we look at the top-10 requesters and communities they are associated with. The top-10 requester graph is depicted by Fig. 2.6. Table 2.2 shows the number of clusters the requester is associated with (3rd column) and the number of tasks (4th column). In addition, the table depicts in the last column the rounded values of the $\mathscr{A}$ scores to show how the scores among top-ranked requesters are distributed.

The requester Smartsheet.com Clients clearly outranks other requesters. It has also posted a large number of tasks to relevant communities. Between the 5th to the 10th ranked requesters one can see less significant differences in the ranking scores because the number of tasks and clusters are also not significantly different. The number one ranked community in AMT using our community discovery and ranking approach is the community dealing with 'data' and 'collection'. Each requester in the top-10 list is associated with a number of communities and all requesters are also connected with the top community.

Next, we filter the graph and show the top-10 communities and the associated requesters in Fig. 2.7 (best viewed online). Descriptions on the top-ranked communities are given in Table 2.3.

The top-ranked community deals with 'data' and 'collection' and can be easily located in the graph by looking at the triangular node which has a dense neighborhood of requesters (top left in Fig. 2.7). Table 2.3 shows in addition to the cluster-based keywords (last column) the total number of tasks found in the community (3rd column), the number of top-10 ranked requesters connected to a given community (4th column), and the number of top-10 ranked requester tasks (also in 4th column in parentheses). One can observe that top-ranked communities have also top-ranked

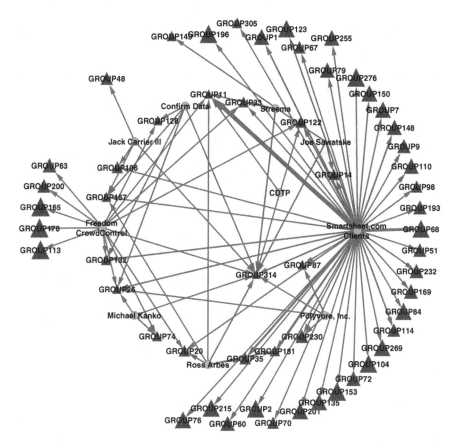

Fig. 2.6 Top-10 requester graph

Table 2.2 Description of top-10 requester graph

Rank	Requester name	Number of clusters	Number of tasks	$\mathscr{A}$ (rounded)
1	Smartsheet.com Clients	46	877	0.002706
2	Freedom CrowdControl	14	51	0.002387
3	Ross Arbes	6	56	0.002269
4	Confirm Data	5	16	0.002249
5	Polyvore, Inc.	4	20	0.002206
6	Streema	3	22	0.002203
7	CDTP	3	13	0.002197
8	Jack Carrier III	4	12	0.002194
9	Michael Kanko	3	14	0.002192
10	Joe Sawatske	3	19	0.002183

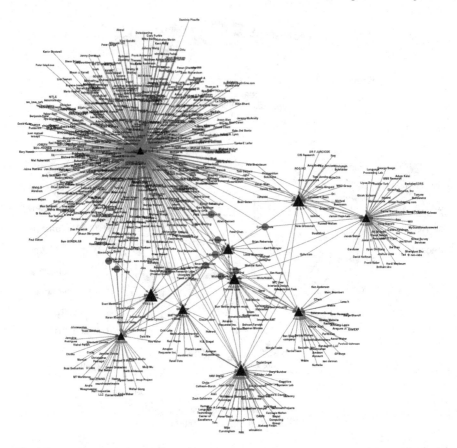

Fig. 2.7 Top-10 community graph

requesters associated with them, which is the desired behavior of our ranking model. Also, since the algorithm takes weighted edges into account, the number of tasks that are actually posted in a community play a key-role.

To conclude our discussions, top-ranked requesters are identified based on their active contribution (posting tasks) to top-ranked communities.

2.6.2 Recommendation of Crowdsourcing Brokers

In this section, we discuss the performed experiments to discovery crowdsourcing brokers. First, we take the entire set V_R and establish the keyword-based profiles of each requester. In the next step, we apply the Algorithm 2 to construct G_{PG} using the matching function as defined in (2.4). To find a suitable threshold ξ, we generated a number of graphs with varying thresholds $0.0 \leq \xi \leq 1.0$ and measured the indegree

Table 2.3 Description of top-10 community graph

Rank	Community Id	Number of tasks	Number top-requesters	Keywords
1	GROUP314	1222	10 (109)	Data, collection
2	GROUP11	801	5 (352)	Actors, website
3	GROUP128	136	2 (4)	Photo, article, social
4	GROUP26	146	5 (20)	TV, web
5	GROUP87	139	2 (4)	Review, shopping
6	GROUP157	34	3 (18)	Marketing, phone
7	GROUP149	53	1 (1)	Moteur, question
8	GROUP74	250	2 (10)	Assistance, text
9	GROUP305	92	1 (4)	Fast, quick
10	GROUP181	28	2 (9)	Interesting, business

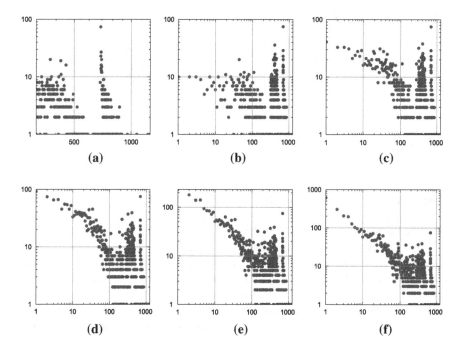

Fig. 2.8 Degree distributions under different connectivity thresholds. **a** $\xi = 0.0$. **b** $\xi = 0.1$. **c** $\xi = 0.2$. **d** $\xi = 0.3$. **e** $\xi = 0.4$. **f** $\xi = 0.5$

of each node in G_{PG}. The following series of scatter plots and shows the indegree distributions for the interval $0.0 \leq \xi \leq 0.5$ (Fig. 2.8) and $0.6 \leq \xi \leq 1.0$ (Fig. 2.9) respectively. On the horizontal axis, the indegree is shown and on the vertical axis the number of nodes. Recall that the set V_R holds 5584 distinct requesters.

The plots in Figs. 2.8a and 2.9e can be regarded as boundaries. However, notice that the profile match *pm* must be greater than ξ (see Algorithm 2). Otherwise each

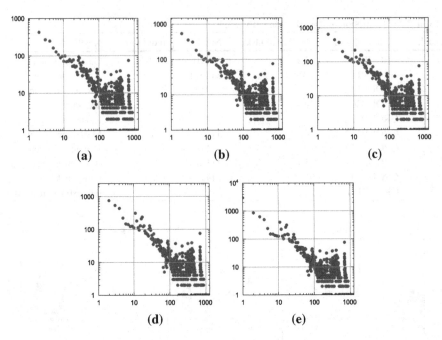

Fig. 2.9 Degree distributions under different connectivity thresholds. **a** $\xi = 0.6$. **b** $\xi = 0.7$. **c** $\xi = 0.8$. **d** $\xi = 0.9$. **e** $\xi = 1.0$

requester would be connected to all other requesters yielding an indegree of 5583 for all requesters. In the case of $pm > 0.0$ (Fig. 2.8a), the indegree is almost evenly distributed with an average degree of 500.

Next we increased ξ and observe that the indegree yields a shape similar to those indegree distributions found in naturally emerging graphs [2]. The majority of nodes has a low indegree whereas a lower number of nodes has a high indegree. This scaling law is also clearly visible when setting a threshold of $\xi > 0.4$. This behavior fits well our proposal where a few requesters would qualify for being brokers that transmit tasks on behalf of others (i.e., the majority of nodes) to crowdsourcing platforms such as AMT or oDesk. Within the interval $\xi = [0.5, 1.0]$ the degree distributions exhibit a similar shape.

In subsequent discussions, we chose a threshold of $\xi = 0.5$ since higher thresholds would not drastically change the shape of the distribution.

The next step in the broker ranking approach is to take the graph G_{PG} and calculate ppr scores using (2.6). To illustrate the approach, we set the query keywords as $Q = \{$'korrigieren', 'deutsch'$\}$ to find and rank requesters which would be suitable brokers for tasks related to correcting German related documents, articles, etc. The personalization vector $p(u; Q)$ was calculated using Algorithm 3. Furthermore, (2.6) was parameterized with $\alpha = 0.15$. The top-10 results are visualized by Fig. 2.10. The node size is based on the position (1–10) in the ranking results where the number 1

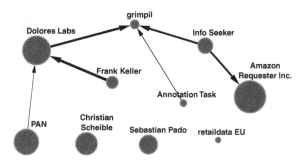

Fig. 2.10 Top-10 ranked requesters in G_{PG}

ranked node has the largest size and the number 10 ranked node the smallest size. The edge width is based on the matching score calculated by using (2.4). Only edges among the top-10 ranked requesters are shown. Information regarding the top-10 ranked requesters is further detailed in Table 2.4.

We made the following important observation when performing broker discovery in G_{PG}. The greatest benefit of applying a network-centric approach to ranking requesters is the discovery of related requesters that may not actually match any of the keywords provided by Q. Suppose the following case where $\alpha = 1$ yielding

$$ppr(u; Q) = p(u; Q), \quad \text{with} \quad \alpha = 1. \tag{2.7}$$

Requesters are assigned a value of 0 if none of their profile keywords match Q and otherwise the weight based on keyword frequency as described in Algorithm 3. In other words, requesters are ranked based on simple keyword-based matching and frequency-based weighting. The results are quite similar by having the top-7 requesters (see Table 2.4) ranked in the same order:

Table 2.4 Description of top-10 ranked requesters in G_{PG}

Rank	Requester name	Indegree	Number of tasks (total)	*ppr* (rounded)
1	Amazon Requester Inc.	25	1021	0.139038
2	Dolores Labs	933	2063	0.006523
3	PAN	12	11	0.002715
4	Christian Scheible	27	7	0.002036
5	Sebastian Pado	1	2	0.001357
6	Info Seeker	4	2	0.001357
7	Frank Keller	16	3	0.000679
8	Grimpil	60	3	0.000577
9	Annotation Task	227	18	0.000245
10	Retaildata EU	43	21	0.000002

Table 2.5 Top-ranked
requesters using (2.7)

Rank	Requester name
1	Amazon Requester Inc.
2	Dolores Labs
3	PAN
4	Christian Scheible
5	Sebastian Pado
6	Info Seeker
7	Frank Keller

Notice, however, only 7 out of 5583 requesters exactly match the query $Q =$ {'korrigieren', 'deutsch'}. Thus, all other nodes will receive a ranking score of 0. By applying our proposed approach using (2.6) we have the following important benefits:

- Since importance of requesters is computed relative to the nodes specified in the personalization vector, *all* nodes (requesters) receive a ranking score.
- Requesters that do not match the query but are connected in G_{PG} with other high ranked requesters will be able to improve their position in the ranking results.

Amazon Requester Inc. is the highest ranked requester in either case, $\alpha = 0.15$ and $\alpha = 1$, with regards to the keywords specified in Q. Therefore, this requester would be the most recommendable broker. However, by using $\alpha = 0.15$ other requesters such as grimpil (see requester with rank 8 in Table 2.4) who has strong inbound links from Dolores Labs and Info Seeker are also discovered in the top-10 list. This requester, for example, would have not been discovered otherwise.

Overall, our approach provides an important tool for the discovery of brokers by establishing a profile-relationship graph G_{PG} and by ranking requesters based on their actual match (i.e., $p(u; Q)$) and based on their degree of connectivity (the second part of (2.6)—that is $\sum_{(v,u) \in E_{PG}} \frac{ppr(v)}{\text{outdegree}(v)}$).

Indeed, both components cannot be computed independently by simply summing them up. Instead, the *pr* and *ppr* scores must by computed in an iterative process that needs to converge towards a fixed value (see [23, 40]). Finally, our approach lets other requesters desiring to utilize brokers for crowdsourcing their tasks to discovery the best matching brokers and also pathways revealed by G_{PG} to matching brokers (e.g., a suitable path to Amazon Requester Inc. can be established via Info Seeker).

As a final remark on the experiments, time complexity of the presented clustering and ranking algorithms becomes an issue as the size of the number of keywords and the number of requesters increases. As mentioned in Sect. 2.4, at this point we have considered a subset of about 300 keywords that have already been filtered. Furthermore, we only used those keywords that had a co-occurrence frequency of at least 10 with some other keyword (minimum threshold). Thus, time complexity has not been an important issues in our currently conducted experiments but deserves attention in our future experiments with larger crowdsourcing datasets.

2.7 Conclusion and Future Work

Crowdsourcing is a new model of outsourcing tasks to Internet-based platforms. Models for community detection and broker discovery have not been provided by existing research. In this work we introduced a novel community discovery and ranking approach for task-based crowdsourcing markets. We analyzed the basic marketplace statistics of AMT and derived a model for clustering tasks and requesters. The presented approach and algorithm deliver very good results and will help to greatly improve the way requesters and workers discover new tasks or topics of interest.

We have motivated and introduced a broker discovery and ranking model that lets other requesters discovery intermediaries who can crowdsource tasks on their behalf. The motivation for this new broker based model can be manifold. As an example, brokers allow large businesses and corporations to crowdsource tasks without having to worry about framing and posting tasks to crowdsourcing marketplaces

In future work we will compare the presented hierarchical clustering approach with other techniques such as Latent Dirichlet Allocation (LDA) [4]. In addition, we will evaluate the quality of the keyword-based cluster as well as the community rankings through crowdsourcing techniques (cf. also [11]). With regards to community evolution, we will analyze the dynamics of communities (birth, expansion, contraction, and death) by looking at the task posting behavior of requesters. This will help to make predictions about the needed number of workers with a particular set of skills. The crowdsourcing platform may manage resource demands by creating training tasks to prevent shortcomings in the availability of workers that satisfy task requirements. Some of the related issues have been tackled in our previous work [48] but an integration with the present work is needed.

Furthermore, based on our broker discovery approach, we will look at different negotiation and service level agreement setup strategies. The personalization vector could be computed based on further parameters such as costs, the requesters availability and reliability constraints. Finally, standardization issues of interfaces towards crowdsourcing platforms in general as well as interfaces for brokers will be part of future research.

References

1. Alonso, O., Rose, D.E., Stewart, B.: Crowdsourcing for relevance evaluation. SIGIR Forum **42**(2), 9–15 (2008)
2. Barabasi, A.-L., Albert, R.: Emergence of scaling in random networks. Science **286**, 509 (1999)
3. Bhattacharyya, P., Garg, A., Wu, S.: Analysis of user keyword similarity in online social networks. Soc. Netw. Anal. Min. **1**, 143–158 (2011). doi:10.1007/s13278-010-0006-4
4. Blei, D.M., Ng, A.Y., Jordan, M.I.: Latent dirichlet allocation. J. Mach. Learn. Res. **3**, 993–1022 (2003)
5. Branting, L.: Context-sensitive detection of local community structure. Soc. Netw. Anal. Min. **1**, 1–11 (2012). doi:10.1007/s13278-011-0035-7

6. Burt, R.S.: Structural Holes: The Social Structure of Competition. Harvard University Press, Cambridge (1992)
7. Callison-Burch, C., Dredze M.: Creating speech and language data with amazon's mechanical turk. In: Proceedings of the NAACL HLT 2010 Workshop on Creating Speech and Language Data with Amazon's Mechanical Turk, CSLDAMT '10. Association for Computational Linguistics, pp. 1–12. Stroudsburg, PA, USA (2010)
8. Carvalho, V.R., Lease, M., Yilmaz, E.: Crowdsourcing for search evaluation. SIGIR Forum **44**(2), 17–22 (2011)
9. Cazabet, R., Takeda, H., Hamasaki, M., Amblard, F.: Using dynamic community detection to identify trends in user-generated content. Soc. Netw. Anal. Min. 1–11 (2012). doi:10.1007/s13278-012-0074-8
10. Chakrabarti, S.: Dynamic personalized pagerank in entity-relation graphs. In: Proceedings of the 16th International Conference on World Wide Web, WWW '07, pp. 571–580. ACM, New York (2007)
11. Chang, J., Boyd-Graber, J., Gerrish, S., Wang, C., Blei, D.: Reading tea leaves: How humans interpret topic models. In: Bengio, Y., Schuurmans, D., Lafferty, J., Williams, C.K.I., Culotta, A. (eds.) Advances in Neural Information Processing Systems 22, pp. 288–296. MIT press, Cambridge (2009)
12. ClickWorker: http://www.clickworker.com/ (2012). Accessed 20 Aug
13. CrowdFlower: http://crowdflower.com/ (2012). Accessed 20 Aug
14. Doan, A., Ramakrishnan, R., Halevy, A.Y.: Crowdsourcing systems on the world-wide web. Commun. ACM **54**(4), 86–96 (2011)
15. Eda, T., Yoshikawa, M., Yamamuro, M.: Locally expandable allocation of folksonomy tags in a directed acyclic graph. In: Proceedings of the 9th International Conference on Web Information Systems Engineering, WISE '08, pp. 151–162. Springer, Berlin, Heidelberg (2008)
16. Fazeen, M., Dantu, R., Guturu, P.: Identification of leaders, lurkers, associates and spammers in a social network: context-dependent and context-independent approaches. Soc. Netw. Anal. Min. **1**, 241–254 (2011). doi:10.1007/s13278-011-0017-9
17. Fisher, D., Smith, M., Welser, H.T.: You are who you talk to: detecting roles in usenet newsgroups. In: Proceedings of the 39th Annual Hawaii International Conference on System Sciences, HICSS '06, Vol. 3, p. 59.2, IEEE Computer Society, Washington, DC, USA (2006)
18. Flickr: http://www.flickr.com/ (2012). Accessed 20 Aug
19. Fogaras, D., Rácz, B., Csalogány, K., Sarlós, T.: Towards scaling fully personalized pagerank: algorithms, lower bounds, and experiments. Internet Math. **2**(3), 333–358 (2005)
20. Franklin, M.J., Kossmann, D., Kraska, T., Ramesh, S., Xin, R.: Crowddb: answering queries with crowdsourcing. In: Proceedings of the 2011 International Conference on Management of Data, SIGMOD '11, pp. 61–72. ACM, New York (2011)
21. Gemmell, J., Shepitsen, A., Mobasher, B., Burke, R.: Personalizing navigation in folksonomies using hierarchical tag clustering. In: Proceedings of the 10th International Conference on Data Warehousing and Knowledge Discovery, DaWaK '08, pp. 196–205. Springer, Berlin, Heidelberg (2008)
22. Golder, S., Huberman, B.A.: Usage patterns of collaborative tagging systems. J. Inform. Sci. **32**(2), 198–208 (2006)
23. Haveliwala, T.H.: Topic-sensitive pagerank. In: Proceedings of the 11th International Conference on World Wide Web, WWW '02, pp. 517–526. ACM, New York (2002)
24. Heer, J., Bostock, M.: Crowdsourcing graphical perception: using mechanical turk to assess visualization design. In: Proceedings of the 28th International Conference on Human factors in Computing Systems, CHI '10, pp. 203–212. ACM, New York (2010)
25. Herlocker, J.L., Konstan, J.A., Terveen, L.G., Riedl, J.T.: Evaluating collaborative filtering recommender systems. ACM Trans. Inf. Syst. **22**(1), 5–53 (2004)
26. Heymann, P., Garcia-Molina, H.: Collaborative creation of communal hierarchical taxonomies in social tagging systems. Technical report, Computer Science Department, Standford University, April (2006)
27. Howe, J.: The rise of crowdsourcing. Wired **14**(14), 1–5 (2006)

28. Howe, J.: Crowdsourcing: Why the Power of the Crowd is Driving the Future of Business. Crown Business, New York (2008)

29. Ipeirotis, P.G.: Analyzing the amazon mechanical turk marketplace. XRDS **17**, 16–21 (2010)

30. Jeh, G., Widom, J.: Scaling personalized web search. In: Proceedings of the 12th International Conference on World Wide Web, WWW '03, pp. 271–279. ACM, New York (2003)

31. Kittur, A., Chi, E.H., Suh, B.: Crowdsourcing user studies with mechanical turk. In: Proceedings of the 26th Annual SIGCHI Conference on Human factors in Computing Systems, CHI '08, pp. 453–456. ACM, New York (2008)

32. Kleinberg, J.M.: Authoritative sources in a hyperlinked environment. J. ACM **46**(5), 604–632 (1999)

33. Kourtellis, N., Alahakoon, T., Simha, R., Iamnitchi, A., Tripathi, R.: Identifying high betweenness centrality nodes in large social networks. Soc. Netw. Anal. Min. 1–16 (2012). doi:10.1007/s13278-012-0076-6

34. Lampe, C., Resnick, P.: Slash(dot) and burn: distributed moderation in a large online conversation space. In: Proceedings of the SIGCHI Conference on Human Factors in Computing Systems, CHI '04, pp. 543–550. ACM, New York (2004)

35. Little, G., Chilton, L.B., Goldman, M., Miller, R.C.: Turkit: human computation algorithms on mechanical turk. In: Proceedings of the 23rd Annual ACM symposium on User Interface Software and Technology, UIST '10, pp. 57–66. ACM, New York (2010)

36. Marge, M., Banerjee, S., Rudnicky, A.I.: Using the amazon mechanical turk for transcription of spoken language. In: Proceedings of the IEEE International Conference on Acoustics, Speech, and Signal Processing, pp. 5270–5273, 2010

37. Michlmayr, E., Cayzer, S.: Learning user profiles from tagging data and leveraging them for personal(ized) information access. In: Tagging and Metadata for Social Information Organization, Workshop, WWW07, 2007

38. Munro, R., Bethard, S., Kuperman, V., Lai, V.T., Melnick, R., Potts,C., Schnoebelen, T., Tily, H.: Crowdsourcing and language studies: the new generation of linguistic data. In: Proceedings of the NAACL HLT 2010 Workshop on Creating Speech and Language Data with Amazon's Mechanical Turk, CSLDAMT '10, pp. 122–130. Association for Computational Linguistics, Stroudsburg, PA, USA (2010)

39. oDesk: http://www.odesk.com/ (2012). Accessed 20 Aug

40. Page, L., Brin, S., Motwani, R., Winograd, T.: Bringing order to the web. The pagerank citation ranking (1999)

41. Parameswaran, A., Park, H., Garcia-Molina, H., Polyzotis, N., Widom, J.: Deco: declarative crowdsourcing. Technical report, Stanford University (2011)

42. Psaier, H., Skopik, F., Schall, D., Dustdar, S.: Resource and agreement management in dynamic crowdcomputing environments. In: EDOC, pp. 193–202. IEEE Computer Society, Washington, DC (2011)

43. Quinn, A.J., Bederson, B.B.: Human computation: a survey and taxonomy of a growing field. In: Proceedings of the 2011 Annual Conference on Human Factors in Computing Systems, CHI '11, pp. 1403–1412. ACM, New York (2011)

44. Romesburg, C.: Cluster Analysis for Researchers. Krieger, Florida (2004)

45. Rosvall, M., Bergstrom, C.T.: Maps of random walks on complex networks reveal community structure. PNAS **105**, 1118 (2008)

46. Salton, G., Buckley, C.: Term-weighting approaches in automatic text retrieval. Inf. Process. Manage. **24**(5), 513–523 (1988)

47. Samasource: http://samasource.org/ (2012). Accessed 20 Aug

48. Satzger, B., Psaier, H., Schall, D., Dustdar, S.: Stimulating skill evolution in market-based crowdsourcing. In: BPM, pp. 66–82. Springer, Berlin (2011)

49. Schall, D.: A human-centric runtime framework for mixed service-oriented systems. Distributed and Parallel Databases, **29**, 333–360 (2011). doi:10.1007/s10619-011-7081-z(Springer, Berlin)

50. Schall, D.: Expertise ranking using activity and contextual link measures. Data Knowl. Eng. **71**(1), 92–113 (2012). doi:10.1016/j.datak.2011.08.001

51. Schall, D., Skopik, F.: An analysis of the structure and dynamics of large-scale q/a communities. In: Eder, J., Bieliková, M., Tjoa, A.M. (eds.) ADBIS, Lecture Notes in Computer Science, vol. 6909, pp. 285–301. Springer, Berlin (2011)
52. Schall, D., Skopik, F., Psaier, H., Dustdar, S.: Bridging socially-enhanced virtual communities. In: Chu, W.C., Wong, W.E., Palakal, M.J., Hung, C.-C. (eds.) SAC, pp. 792–799. ACM, New York (2011)
53. Shepitsen, A., Gemmell, J., Mobasher, B., Burke, R.: Personalized recommendation in social tagging systems using hierarchical clustering. In: Proceedings of the 2008 ACM Conference on Recommender Systems, RecSys'08, pp. 259–266. ACM, New York (2008)
54. Sigurbjörnsson, B., van Zwol, R.: Flickr tag recommendation based on collective knowledge. In: Proceedings of the 17th International Conference on World Wide Web, WWW'08, pp. 327–336. ACM, New York (2008)
55. Skopik, F., Schall, D., Dustdar, S.: Start trusting strangers? bootstrapping and prediction of trust. In: Vossen, G., Long, D.D.E., Yu, J.X. (eds.) WISE, Lecture Notes in Computer Science, vol. 5802, pp. 275–289. Springer, Berlin (2009)
56. SmartSheet. http://www.smartsheet.com/ (2012). Accessed 20 Aug
57. SpeechInk. http://www.speechink.com/ (2012). Accessed 20 Aug
58. Vukovic, M.: Crowdsourcing for enterprises. In: Proceedings of the 2009 Congress on Services-I, SERVICES '09, pp. 686–692. IEEE Computer Society, Washington, DC (2009)

Chapter 3
Human-Provided Services

Abstract In this chapter, we discuss collaboration scenarios where people define services based on their dynamically changing skills and expertise by using Human-Provided Services. This approach is motivated by the need to support novel service-oriented applications in emerging crowdsourcing environments. In such open and dynamic environments, user participation is often driven by intrinsic incentives and actors properties such as reputation. We present a framework enabling users to define personal services to cope with complex interactions. We focus on the discovery and provisioning of human expertise in service-oriented environments.

Keywords Human provided services · Mixed service-oriented systems · Crowd-sourcing · Social computing

3.1 Introduction

The transformation of how people collaborate and interact on the Web has been poorly leveraged in existing SOA. In SOA, compositions are based on Web services following the loose coupling and dynamic discovery paradigm. We argue that people should be able to define interaction interfaces (services) following the same principles to avoid the need for parallel systems of humans and software services. We introduce *mixed service-oriented systems* [26] that are composed of both Software-Based Services (SBS) and Human-Provided Services (HPS) [30], interacting to perform certain activities. Here, user-provided services are well-defined interfaces to interact with people. The problem is that current systems lack the notion of human capabilities in SOA. The challenge is to support the user in providing services in open Web-based environments. HPSs can be discovered in a manner similar to SBS. Following this approach, humans are able to offer HPSs and manage interactions in dynamic collaboration environments.

D. Schall, *Service-Oriented Crowdsourcing*, SpringerBriefs in Computer Science, DOI: 10.1007/978-1-4614-5956-9_3, © The Author(s) 2012

Unlike traditional process-centric environments in SOA, we focus on flexible and open collaboration scenarios. In this chapter, we present the following novel key contributions:

- People need to be able to be able to provide services and to manage interactions in service-oriented systems. We present the HPS architecture and its core components: a *Middleware Layer* providing features for managing data collections and XML artifacts, the *API Layer* comprising services for user forms generation and XSD transformations, a *Runtime Layer* enabling basic activity and user management features as well as support for interactions using Web services technology.
- In open and dynamic environments, expertise profiles need to be maintained in an automated manner to avoid outdated information. We introduce a context-sensitive expertise ranking approach based on interaction mining techniques.
- We evaluate our approach by discussing results of our expertise mining approach.

This chapter is organized as follows. We overview related work in Sect. 3.2. In Sect. 3.3, we present the HPS activity and task model enabling dynamic interactions in service-oriented systems. In Sect. 3.4, we discuss the Human-Provided Services architecture and framework. The discovery and selection of HPS is strongly influenced by human expertise. Our expertise ranking approach based on interaction mining techniques is presented in Sect. 3.5. Section 3.6 presents experiments and implementation details. We conclude the chapter in Sect. 3.7.

3.2 Background

We structure our discussion regarding related work in three topics: (i) *crowdsourcing* to clearly motivate the problem context of our work, (ii) *interaction modeling* to overview different techniques for structuring collaborations, and (iii) *metrics and expertise mining* to track user interest and skills in open Web-based platforms. Our work is specifically based on the assumption that evolving skills and expertise influence how interactions are performed (for example, delegations) in crowdsourcing environments.

Crowdsourcing. In recent years, there has been a growing interest in the complex 'connectedness' of today's society. Phenomena in our online-society involve networks, incentives, and the aggregate behavior of groups [9]. *Human computation* is motivated by the need to outsource certain steps in a computational process to humans [11]. An application of human computation in genetic algorithms was presented in [17]. A variant of human computation called games that matter was introduced by [37]. Related to human computation are systems such as Amazon Mechanical Turk[1] (AMT). AMT is a Web-based, task-centric platform. Users can publish, claim, and process tasks. For example [35], evaluated the task properties of a similar platform in cases where large amounts of data are reviewed by humans. In

---

[1] http://www.mturk.com/

contrast to common question/answer (Q/A) forums, for example Yahoo! Answers,[2] AMT enables businesses to access the manpower of thousands of people using a Web services API. Mixed service-oriented systems [26] target flexible interactions and compositions of Human-Provided and Software-Based Services [30]. This approach is aligned with the vision of the Web 2.0, where people can actively contribute services. In such networks, humans may participate and provide services in a uniform way by using the HPS framework [26]. A similar vision is shared by [23] who defines *emergent collectives* which are networks of interlinked valued nodes (services). In such collectives, there is an easy way to add nodes by distributed actors so that the network will scale. Current crowdsourcing platforms do not support complex interactions (e.g., delegation flows) that require joint capabilities of human and software services.

Questions include: how can people control flexible interaction flows in emerging crowdsourcing environments?

Interaction Modeling. In business processes (typically *closed* environments), human-based process activities and human tasks can be modeled in a standardized service-oriented manner. WS-HumanTask [4] (WS-HT) and BPEL4People [3] (B4P) are related industry standards released to address the need for human involvement in service-oriented systems. These standards and related efforts specify languages to model human interactions in BPEL [3], the lifecycle of humans tasks [4] in SOA, resource patterns [25], and role-based access models [19]. A concrete implementation of B4P as a service was introduced in [36]. A *top-down* approach, however, demands for the precise definition of roles and interactions between humans and services. The application of such models is therefore limited in crowdsourcing due to the complexity of human tasks, people's individual understanding, and unpredictable events. Other approaches focus on ad-hoc workflows or self-contained subprocesses (worklets) [1] based on *activity theory*, and task-adaptation [10] to cope with changing environmental conditions. In [20], business activity patterns were introduced to design flexible applications.

Questions include: how can one control interactions in open and dynamic environments that are governed by the emergence of social preferences, skills and reputation?

Metrics and Expertise Mining. Human tasks metrics in workflow management system have been discussed in [18]. A formal approach to modeling and measuring inconsistencies and deviations, generalized for human-centered systems, was presented in [7]. Studies on distributed teams focus on human performance and interactions [5, 22], as well as in *Enterprise 2.0* environments [6]. Models and algorithms to determine the expertise of users are important in future service-oriented environments [27]. Task-based platforms allow users to share their expertise [38]; or users offer their expertise by helping other users in forums or answer communities [2]. By analyzing email conversations [8], the authors studied graph-based algorithms such as *Hyperlink-Induced Topic Search* [16] and PageRank [21] to estimate the expertise of users. The authors in [32] used a graph-entropy model to measure the importance of users. In [39] the authors applied PageRank in online communities such as Java Q/A

---

[2] http://answers.yahoo.com/

forums. Approaches for calculating personalized PageRank scores were introduced in [13, 14] to enable topic-sensitive queries in search engines, but have not been applied to interaction analysis (social networks). Most existing link-based expertise mining techniques do not consider information related to the *interaction context*.

Questions include: how can interaction mining algorithms track users' expertise, interest, and skills in an automated manner considering context information?

3.3 HPS Interaction Model

The availability of interaction models in open, Web-based platforms such as the motivating crowdsourcing scenario is currently limited. Most existing crowdsourcing platforms do not support interactions and collaborations between users (tasks are typically assigned to individuals). Other Web-based tools (e.g., bulletin boards, email, instant messaging) lack the ability to compose the capabilities of human- and software-based services, which typically requires standardized (XML-based) message formats, interfaces, etc. The main purpose of the proposed interaction model is:

1. Provisioning of human expertise in a service-oriented manner using SOA principles (Sect. 3.3.1). Examples are 'document review' [30] or 'document translation' [31] services provided by human actors.
2. Model to support flexible interactions between crowd members (Sect. 3.3.2).
3. Task model that can be used to link PFL (Process Flow) artifacts (task descriptions) to flexible crowd-activities which are provisioned through HPS (Sect. 3.3.3).

3.3.1 HPS Activity Model

Activities are used for different purposes. People use activities to structure collaborations in a flexible manner. Also, activities enable users to define Human-Provided Services. We now turn to the activity model enabling the use of HPS in various interaction scenarios, for example CFLs (Crowd Flows). The presented activity model in Fig. 3.1 depicts the most important elements to support basic interaction scenarios.

- An *ActivityDeclaration* defines the name and description of an activity, URI, and a set of tags that can be applied to the declaration. Tags are applied by users to associate keywords to declarations.
- The *HPS Interface* relates to an *ActivityDeclaration*. *Name* in the *HPSInterface* depicts the HPSs name, for example, a *review service*. The *HPSInterface* (description) is very similar to the description of conventional SBS. We perform a simple mapping to depict declarations as Web service descriptions (e.g., using WSDL).
- An *HPSGroup* defines the set of people providing a certain type of service established as the relation between *User*, *HPSInterface*, and *HPSGroup*. An

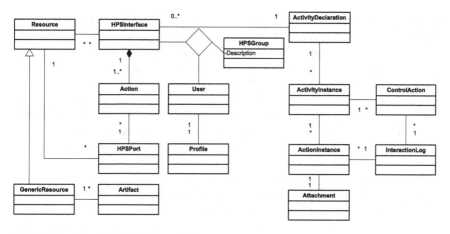

Fig. 3.1 Overview of HPS activity model

HPS requester can be a human seeking the opinion of experts or a composed software service (PFL) requiring human input. A ranking procedure must be used to select the best available HPS. We term the relation between *User* and *HPSInterface personal service*, which is technically an instance of an HPS. Each user has a *Profile* identifying people and storing user preferences.

- A *Resource* is used for different purposes. As mentioned before, *HPSInterfaces* are depicted using languages such as WSDL. Thus, the interface is an XML document that can be modified by using resource identifiers (URIs) to retrieve or update resources. Other resources are *type definitions*, for example, activity types and/or parts of complex data types.
- A *GenericResource* is a special type of *Resource*, which we use to wrap *Artifacts*. *Artifacts* include collaboration documents and all sorts of files that are used and created during collaborations. The *GenericResource* defines metadata associated with *Artifacts*.
- The *Action* concept is used to interact with HPSs in the scope of an activity. The *HPSInterface* is composed of a set of *Actions*. Notice, there are different action concepts in our model. On the one hand, *Action*, as discussed here, is defined by the user in the scope of an *HPSInterface*. The definition of an *Action* is done at *design time*.
- The *HPSPort* depicts the technical—in a Web services sense—realization of an HPS interface. (The details are not needed at this point and will be discussed in the HPS framework section.) The *HPSPort* relates to a set of resources (e.g., typed messages), which are used in certain *Actions*.

The previous concepts were introduced as models to depict and design HPSs. The following concepts describe activity and HPS-centric interactions at *run-time*.

- An *ActivityInstance* represents an actual work item. An activity can be performed many times, which are called instances of the activity. Each instance corresponds to a declaration. Instances represent the context of interactions.

- An *ActionInstance* is connected to an *ActivityInstance*. Each *ActionInstance* is defined by an *Action*. An *Attachment* is something generic to associate XML documents, for example, XML messages that are exchanged between services, and other content-types with an *ActionInstance*. *Attachments* usually convey typed messages that are defined in an *HPSInterface* and *Resources*.

Both *ControlAction* and *ActionInstance* are used at run-time. A *ControlAction*, however, depicts common action types in human collaboration. *ControlActions* include *coordination*, *communication*, and *execution* actions that are associated with instances of activities, for example *delegations* of activities. However, such actions are not part of an *HPSInterface*.

A *ControlAction* is always used between two or more people to, for example, coordinate the execution of activities; whereas an *ActionInstance* may be the result of interactions between human and software services. Each action, *ControlAction* as well as *ActionInstance*, is logged to keep a history of interactions. The *InteractionLog* captures traces of interactions (activities and their actions) performed in collaborations. Also, interactions between software services are logged to maintain a history of the collaboration context.

3.3.2 Hierarchical Activities

Activities can be structured as hierarchies (see Fig. 3.2) using *parent* and *child* relations. Child activities specify the details with respect to the (sub-)steps in collaborations, for example, sub-activities in the scope of a parent activity. This allows for the refinement of collaboration structures as the demand for a new set of activities (e.g., performed by different people and services) increases. The need for the dynamic refinement of collaboration structures is especially required when people have limited experience (the history of performed activities) with respect to a given objective or goal. Furthermore, some people tend to underestimate the scale and complexity of an activity; thus the hierarchical model enables the segmentation of activities into sub-activities, which can be, for example, delegated to other people.

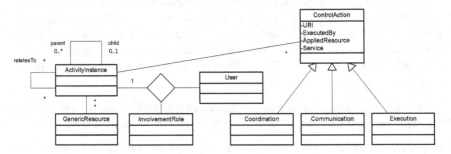

Fig. 3.2 Excerpt of hierarchical activity model

The basic HPS activity model (cf. Fig. 3.1) did not define any notion of activity hierarchies because, currently, we do not support the mapping of activity hierarchies onto HPSs. For example, hierarchically structured activities in activity declarations would require a mapping of such hierarchies into a set of *Actions*. Activities have a *relatedTo* property which provides a mechanism to link to any other activity. Typically, multiple members work on the same activity with different roles. The *InvolvementRole* identifies the creator, observer, contributor, responsible, and supervisor of an activity. Involved workers apply a set of *GenericResources* to perform their work. As mentioned before, objects such as documents are represented as a shared *Artifact*. A *ControlAction* captures activity-change events, interactions between members, and work carried out. Actions can trigger events describing the progress of activities.

3.3.3 Task Model

In most collaborations, activities need to be controlled by capturing temporal aspects such as *progress* of activities and monitoring of *deadlines*. In this section, we define an extended task model, which can be used in open collaboration scenarios; for example, in HPS-based collaborations on the Web. Figure 3.3 shows task-related concepts and their relation to previously introduced concepts.

Controlling the execution of activities. The most fundamental aspect is to control the execution of activities by associating a *HumanTask* with an *ActivityInstance*. Multiple tasks can be created because activity instances can be divided into sub-activities. A *HumanTask* is derived from a generic *Task* defining basic task-properties—*StartAt*, *DueAt*, and, *Priority*. If tasks are used in HPS-based collaborations, requesters are aware of the state of a given interaction (e.g., *accepted*, *inprogress*, or *completed*). Based on these execution parameters, for example, the properties *Priority* and *DueAt*, *Notifications* can be sent to a set of people. Examples include, notify a set of people (*PeopleGroup*) about the status of an activity, escalate deviations in the execution of activities, or notify the supervisor of an activity when the activity (or one of its sub-activities) has been completed.

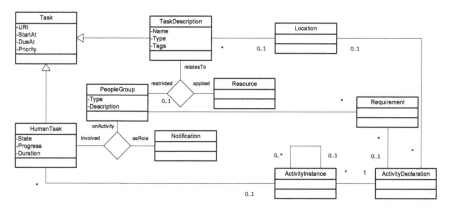

Fig. 3.3 Overview task model

This model is well aligned with the WS-HumanTask [4] (WS-HT) specification. Moreover, functional properties can be associated with *ActivityDeclarations*, depicted as *Requirement* in Fig. 3.3; for example, role models controlling whether users are allowed to work on activities. Again, a generic *PeopleGroup* is used which is populated with a set of people depending on specified requirement. Notice, requirements typically do not change over time. For example, if we use a role model to control the set of people who can work on an activity, we follow a *top-down* view—modeling how an activity should be performed. In contrast, *constraints* change over time depending on the run-time context. Constraint are, for example, the minimum set of skills or level of expertise a potential worker must have. Indeed, skills and level of expertise change over time depending on performed activities.

Creating announcements. The idea of the HPS model is not only to support enterprise collaboration scenarios but also Web-based crowd collaborations. In enterprises, a corporate directory usually holds all information regarding employees, their role in the company, and additional contact information, which can be accessed to populate a *PeopleGroup*. However, these announcements are well applicable to enterprise collaborations as well because in global corporations it is impossible to maintain expertise, roles, interests of employees in a central directory.

Here we define two scenarios showing the usefulness of announcements:

- We can imagine a *TaskDescription* as an announcement to express the need for a set of HPSs to work on tasks. The notion of task descriptions is similar to marketplaces of work in task-based platforms on the Web, for example, Amazon's Mechanical Turk where Human Intelligence Tasks (HITs) are used for this purpose. See the relation between *TaskDescription*, *Resource*, and *PeopleGroup* in Fig. 3.3. A *Resource* describes an HPS as previously discussed in the basic HPS activity model.

  Task descriptions comprise constraints such as task availability information (beginning and expiration time of the task) and the number of available task instances (how many of those tasks can be claimed by users). In this case, it is clear that a particular type of HPS has to be used in the context of a task.

- The relation between *ActivityDeclaration*, *TaskDescription*, and *Location* depicts the need for a service—potentially in a specific location area.

  Therefore, these kind of announcements are opportunities for users to create *new* HPSs or to associate an existing HPS with a description which has not been considered before. Such announcements are different with respect to the previous case (marketplace example) because *ActivityDeclaration* and *TaskDescription* do not demand for a particular type of HPS.

3.3.4 Task Execution Model

The next step is to introduce a *task execution model* defining the possible task states. The task execution model is depicted by Fig. 3.4. It is relevant for both cases, announcements of task and the control of activity executions.

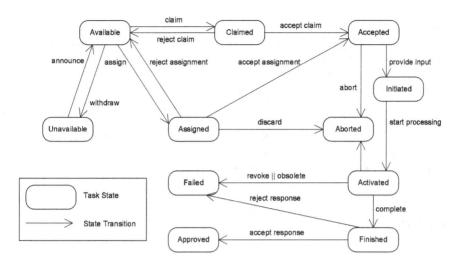

Fig. 3.4 HPS Task execution model

- *Claiming Tasks*: Announcements allow requesters to denote the availability of work items (i.e., activities) without explicitly selecting a particular HPS. Announcement can be generated if there is not any matching HPS available; or if the demanded HPS is currently not provided by users. Initially, a task is set to *Available* and becomes *Unavailable* when the announcement expires. Based on announcements, tasks can be *Claimed*, *Accepted* or rejected by requesters (becoming *Available* again).
- *Task Assignments*: A task can be *Assigned* to HPSs without issuing announcements, specifically when software services generate tasks that need to be processed based on, for example, deadlines. An *Assigned* task may go into the *Accepted* state, otherwise to *Aborted* when the assignment procedure times out. For example, the user is not responding to an assignment request.

The task state changes from *Accepted* to *Initiated* when an action is performed in the context of an activity (e.g., sending a request to an HPS). The task changes its state to *Aborted* if the initiation fails (*Initiated* state). The state *Activated* indicates that the request is processed, followed by the *Finished* state or *Failed* if the HPS was unable to deliver the desired output—the expected information, which can be validated by, for example, a (human) requester reviewing the output. A task is successful if the output of an HPS is *Approved* by the requester.

3.4 Architecture

Most systems based on SOA-principles (registration, discovery, and interaction) typically lack the notion of human capabilities that can be provisioned as service. Traditional service-oriented systems provide support for software-based services only. We propose *mixed service-oriented systems*. Moreover, existing tools for designing services usually require 'expert' knowledge in terms of understanding various WS-standards. HPS harnesses human capabilities within service-oriented environments while leveraging Web 2.0 innovations.

In addition, our architecture provides an approach for interaction monitoring that captures the *context of interactions* through activity identifiers. Monitored interactions are logged and analyzed to calculate various metrics such as reputation and interest profiles of users. The automatic calculation of reputation and skills through mining is a novel technique to support the discovery of human capabilities in SOA considering changing expertise and interests of people.

3.4.1 HPS Framework

This section details the HPS framework by introducing various services to enable human interactions in SOA. The HPS platform allows requesters (people and software services) to find and interact with HPSs. The framework offers a set of tools to support the design of HPSs and a middleware hosting various services such as the HPS Access Layer (HAL). Figure 3.5 shows the main components of the HPS framework. The arrows in the figure depict a 'using' relation between various blocks.

Design Tools. HPS Design tools allow users to create service interfaces (annotation 1) in a simplified manner. These tools are hosted in a Web portal (see [26] for details). Figure 3.5 illustrates the design flow:

- *Interface and Message Formats*: the HPS framework provides tools to automatically translate high level specifications (e.g., activities and interface elements) into low level service descriptions (annotation 2) without requiring the user to understand underlying technologies such as XML or WSDL.
- *Publication of Design Artifacts*: artifacts such as message formats and activity definitions are saved in XML collections (annotation 3).

API Layer. The framework includes services and tools for the design of HPS as well as runtime support for the automatic generation of interfaces. The *API Layer* includes the following core services:

- *WSDL API* service to generate service descriptions; in particular, to create WSDLs based on human activities and user specified interface elements (parameters and complex elements)
- *Forms API* implementing support for XML Forms (XForms)
- *XSD Transformer* service utilizing the *Forms API* to automatically generate XForms based on XML schema definitions, for example, as defined in WSDL documents
- *Tag Management* service associating tags with HPS artifacts (activities, actions, and WSDLs)

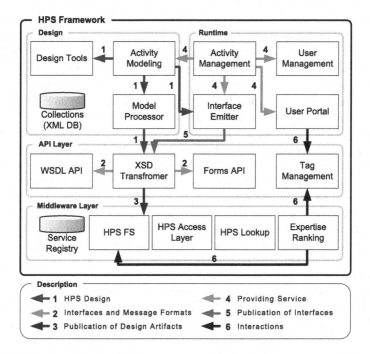

Fig. 3.5 HPS framework and architecture

Runtime Infrastructure Services. The following services have been designed and implemented to enable HPS-based collaboration.

- The *Activity Management* service maintains activity declarations and activity instances (annotation 4).
- The *User Management* service holds data related to profiles and contact details.
- The *Interface Emitter* generates HPS interfaces depending on the interaction scenario (annotation 5); for example, interactions between humans or interactions requiring WSDL interfaces (e.g., compositions of HPS and software services). Since collaboration scenarios include enterprise collaborations, for example, Web-based portals implementing rich user interfaces, and also mobile collaboration scenarios, interface generation can be customized based on the user's current context. Therefore, based on the requirements and constraints of the current or preferred user device, different interface representations can be generated.

Middleware Layer. The *HPS FS* is an XML based, distributed file system to manage user profiles, human tasks, service related information such as WSDL descriptions, and personal services. The HPS FS offers a set of APIs to manage XML artifacts and collections via the Atom Protocol Model[3] to retrieve and update HPS related information. We embed HPS interfaces, described using WSDL, as elements in Atom-based

---

[3] http://tools.ietf.org/html/rfc5023

XML documents (see Atom Syndication Format[4]). Atom-formatted representations contain HPS 'information items' with the advantage that various Web 2.0 authoring tools and APIs can be used to retrieve and update Atom-based elements. HPS information includes: (i) which services are registered with the HPS framework, (ii) how to interact with services, (iii) the geographic location of services; if location information is shared by the user, and (iv) other context information of an HPS including the current availability of a particular service.

HAL dispatches and routes SOAP requests to the corresponding service. Thus, humans and software services (i.e., HPS requesters) are able to interact with HPSs by issuing requests toward the HPS middleware. HAL implements security features to prevent unauthorized access and allows requests to be routed according to user-defined rules (e.g., automatic delegations based on load-conditions [34]). The *HPS Ranking* algorithms are used for the analyses of human and service interactions to recommend the most suitable HPS based on various interaction and task metrics. Ranking results and recommendations can be requested from a *Expertise Ranking* service (annotation 6). The *HPS Lookup* supports various ways to discover HPSs. Web browsers can be used to obtain a list of services as 'news items' embedded in Atom elements. For example, the middleware implements a service which returns XML documents as news feeds containing HPS-related information. We have implemented this mechanism to support the integration of HPS with other Web 2.0 platforms. Also, a Web services-based API can be used to support typical lookup operations to get a list of available services. The middleware hosts a *Service Registry* that is used when the lookup is performed.

3.4.2 Data Collections

The HPS framework utilizes Web services technology to enable HPS at the technical level. Therefore, various XML-based collections and resources need to be managed in an efficient manner. In HPS, XML-based collections are managed by the HPS FS. Basic create, read, update, and delete (CRUD) operations can be performed on HPS-related information. As mentioned before, the Atom protocol is used for this purpose. Resources and collections include:

- *User Profile and Metrics*: Profiles contain *hard* and *soft-facts*. Hard-facts includes information as found in resumes such as education, employment history including organizational information and position held by the user, and professional activities. Soft-facts are represented as competencies. A competency consists of *weights* (skill level of a user), *classification* (description of area or link to taxonomy), and *evidence* (external sources acting as reference or recommendation). Soft-facts can be generated by the middleware based on users' activities to indicate expertise or

---

[4] http://tools.ietf.org/html/rfc4287

skill level. We use friend-of-a-friend (FOAF[5]) profiles to manage social networks structures (e.g., buddy lists) and other user information.

- *Service Registry*: The registry maintains a number of XML documents describing HPS. This information includes a set of service definitions, the list of available services, and information regarding personal services. The term *personal service* was introduced as a metaphor for a service instance. Service instance is a purely technical term to denote the number of physically deployed services that have the same (syntactic) interface characteristics.

- *Task Registry*: Manages human tasks that can be either public tasks (i.e., announcements used to advertise the need for HPSs) or private tasks that are associated with HPS-based interactions to control the status of collaborations. Public tasks are associated with an interaction upon claiming and processing tasks.

3.4.3 Interactions and Monitoring

The HPS framework dynamically generates interfaces for the discovery of services and interactions with users. Next, we show a (simplified) WSDL-based interface description to realize HPS-based *support services* (as introduced in the crowdsourcing scenario).

```
 1  <?xml version="1.0"?>
 2  <wsdl:definitions name="SupportService" ...>
 3  <wsdl:types>
 4   <xsd:schema targetNamespace="http://danielschall.at/rfs">
 5    <xsd:complexType name="GenericResource">
 6    <xsd:sequence>
 7     <xsd:element name="Location" type="xsd:anyURI" />
 8     <xsd:element name="Expires" type="xsd:dateTime" />
 9    <xsd:sequence>
10    </xsd:complexType>
11    <xsd:complexType name="Request">
12    <xsd:sequence>
13     <xsd:element name="SupportResource" type="GenericResource" />
14     <xsd:element name="Comments" type="xsd:string" />
15    </xsd:sequence>
16    </xsd:complexType>
17    <!-- further types ... -->
18    <xsd:element name="SupportRequest" type="Request" />
19    <xsd:element name="AckSupportRequest" type="xsd:string" />
20    <xsd:element name="GetSupportReply" type="xsd:string" />
21    <xsd:element name="SupportReply" type="Reply" />
22   </xsd:schema>
23  <wsdl:types>
24  <wsdl:message name="GetSupport">
25   <wsdl:part name="part1" element="SupportRequest" />
26  </wsdl:message>
27  <wsdl:message name="AckSupportRequest">
28   <wsdl:part name="part1" element="AckSupportRequest" />
29  </wsdl:message>
30  <!-- further messages ... -->
```

---

[5] http://xmlns.com/foaf/spec/

```
31 <wsdl:portType name="HPSSupportPortType">
32 <wsdl:operation name="GetSupport">
33 <wsdl:input xmlns:wsaw="http://.../addressing/wsdl"
34 message="GetSupport" wsaw:Action="urn:GetSupport" >
35 </wsdl:input>
36 <wsdl:output message="AckSupportRequest" />
37 </wsdl:operation>
38 </wsdl:portType>
39 <wsdl:binding name="HALSOAPBinding" type="HPSSupportPortType">
40 <soap:binding style="document"
41 transport="http://xmlsoap.org/soap/http" />
42 </wsdl:binding>
43 </wsdl:definitions>
```

Listing 3.1 HPS WSDL definition

Listing 3.1 shows a complete HPS WSDL example to support the discovery of HPS interfaces. Lines 4–23 define XML type definitions including `GenericResource` and `SupportRequest`. The user can create such definitions by using tools hosted by the HPS platform. In this simplified example, the activity to be performed by a human is the previously mentioned request for support (RFS) activity comprising resources, the actual request, and the reply, which is a complex XML data structure (abbreviated in this example). Lines 24–29 show an excerpt of WSDL messages. However, we only show the request denoted as `SupportRequest`.

The `HPSSupportPortType` is described by lines 31–38. Notice, the HPS Access Layer (HAL) dispatches all interactions. At run-time, HAL extracts and routes messages to the demanded HPS. Since every interaction is entirely asynchronous, interactions (session) identifier are automatically generated by HAL (e.g., `AckSupportRequest`). Finally, lines 39–42 show the `HALSOAPBinding` of the `HPSSupportPortType`.

```
1  <?xml version="1.0"?>
2  <soap:Envelope xmlns:soap="http://www.w3.org/2001/12/soap-envelope"
3  xmlns:wsa="http://schemas.xmlsoap.org/ws/2004/08/addressing"
4  xmlns:hps="http://danielschall.at/"
5  xmlns:types="http://danielschall.at/types"
6  xmlns:rfs="http://danielschall.at/rfs">
7  <soap:Header>
8  <types:timestamp value="2010-03-05"/>
9  <types:delegation hops="3" deadline="2010-03-06"/>
10 <types:activity url="http://.../Activity#42"/>
11 <wsa:MessageID>uuid</wsa:MessageID>
12 <wsa:From>http://.../Actor#Florian</wsa:From>
13 <wsa:ReplyTo>http://.../Actor#Florian</wsa:ReplyTo>
14 <wsa:To>http://.../Actor#Daniel</wsa:To>
15 <wsa:Action>http://.../Type/RFS</wsa:Action>
16 </soap:Header>
17 <soap:Body>
18 <hps:Request>
19 <rfs:subject>WSDL consumption with Axis2</rfs:subject>
20 <rfs:requ>Axis2 reports a parsing error while consuming
21     the given resource. What is wrong?</rfs:requ>
22 <rfs:comments>Used Axis2 1.4</rfs:comments>
23 <rfs:keywords>WSDL, Axis2</rfs:keywords>
24 <rfs:category>Software/SE/General/SE for Internet projects
25 </rfs:category>
26 <rfs:resource>
```

```
27    <!-- details omitted -->
28    </rfs:resource>
29    </hps:Request>
30    </soap:Body>
31    </soap:Envelope>
```

Listing 3.2 Simplified RFS via SOAP example

The HPS Access Layer logs each service interaction (request and response message) through a logging service. RFSs and their responses, exchanged between crowd members, are modeled as traditional SOAP calls, but with header extensions, as shown in Listing 3.2. The most important SOAP-header extensions include: The Timestamp captures the actual creation of the message and is used to calculate temporal interaction metrics, such as the average response time. The tag Delegation holds parameters that influence delegation behavior, such as the number of subsequent delegations numHops (to avoid circulating RFSs) and deadlines. The Activity uri describes the context of interactions that is based on the previously introduced activity model. The MessageID enables message correlation to match request/response pairs. WS-Addressing tags, besides MessageID, are used to route RFSs through the crowd.

Interactions are periodically analyzed to calculate metrics such as reputation and trust between community members. While the depicted architecture follows a centralized approach, the logging facilities are replicated for scalability reasons, and monitoring takes place in a distributed form. Interactions are purged in predefined time intervals, depending on the required depth of history needed by metric calculation plugins (e.g., for trust inference [33]).

3.5 Expertise Ranking

Evolving skills, interests and expertise need to be maintained in an automated manner to avoid outdated profile information. Top-down approaches define interest and expertise areas using taxonomies and ontologies. Here we follow a interaction mining approach that addresses inherent dynamics of flexible collaboration environments.

3.5.1 Context-Sensitive Interaction Mining

Our expertise ranking approach is based on observed interactions (from logs) and analysis of the structure and dynamics of interaction networks. Therefore, an interaction network (see Fig. 3.6a) is modeled as a graph $G = (V, E)$ composed of the set of vertices V and the set of edges E. Note, here the terms edge and link have the same meaning.

We argue that context information is essential for expertise mining. The context of an interaction can be captured by, for example, extracting relevant keywords

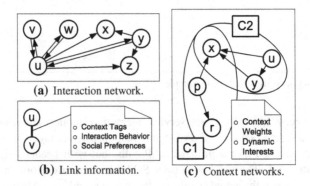

(a) Interaction network.

(b) Link information.   (c) Context networks.

Fig. 3.6 Collaborative networks: **a** interactions are performed between nodes in the network; **b** Metadata and metrics are associated with links between nodes; **c** Context networks are created based in link information

from messages exchanged between users or by tags applied to various collaboration artifacts. In this work, we focus on *tags* (Fig. 3.6b) serving as input for contextual link information. Interactions such as delegation requests are tagged with keywords. As delegation receivers process tasks, our system is able to learn how well people cope with certain tagged tasks; and therefore, able to determine their centers of expertise. The profile $P(u) = \langle f_u(t_1), f_u(t_2), f_u(t_3) \ldots \rangle$ describes the frequencies f_u of tags $T = \{t_1, t_2, t_3 \ldots\}$ that are applied in collaborations by and with u. Interaction metrics such as weights depicting the interest and focus of a user to collaborate with other peers in a specific context are automatically calculated through mining. Figure 3.6c shows networks for context $C1$ and $C2$. Each context network may have one or more tags associated with it.

Existing work in the area of expertise mining (e.g., [39]) typically focuses on a graph representation as depicted by Fig. 3.6a. In contrast, we present an approach and algorithm that is suitable for scenarios as shown in Fig. 3.6c. We base our expertise mining algorithm on well proven and theoretically sound techniques (i.e., see [16, 21]). Specifically, we take the notion of *hubs and authorities* as introduced by Kleinberg [16] as a starting point to derive a context-sensitive expertise mining approach.

3.5.2 Hubs and Authorities

The notion of *authorities* in social or collaborative networks can be interpreted as a measure to estimate the relative standing or *importance* of individuals in social networks. Applying this idea in our crowdsourcing scenario (see Fig. 3.7), a member of the Expert Crowd may receive an RFS and delegate work to some other peer in the network (characterizing hubs in the network). For example (as depicted in Fig. 3.7), u delegates the received RFS to w. Receivers of the delegated work, however, expect

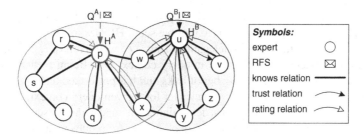

Fig. 3.7 Hubs in different personalized expert queries

RFSs fitting their skills and expertise (i.e., being an authority in the given domain). Careless delegations of work will overload these peers resulting in degraded processing time due to missing expertise.

Within the Expert Crowd, authorities give feedback using rating mechanism (e.g., a number on the scale from 1 to 5) to indicate their satisfaction; i.e., whether a particular hub distributes work according to their skills and interest. Thus, a 'good hub' is characterized by a neighborhood of peers that are satisfied with received RFSs. Also, delegation of work is strongly influenced by *trust*, for example, whether the initial receiver of the RFS (hub within the Expert Crowd) expects that a peer will process work in a reliable and timely manner. RFS receivers need to be trusted by influential hubs that are highly rated to be recognized as authoritative peers in the Expert Crowd.

3.5.3 Personalized Expert Queries

Here we utilize the concept of *personalized expert queries* (as introduce in [27]) to discover *expert hubs* that are well-embedded in expertise networks given a particular query context. A hub is thereby characterized by the social network structure (node degree) *and* connection strength (e.g., count of delegated or processed RFSs) based on joint collaborations. Delegation is important in flexible, interaction-based systems because expert hubs typically attract a large amount of RFSs over time (due to their distinguished expertise). From a network perspective, this means that hubs will be 'bottlenecks' due to the limited capacity and processing speed of the HPS. However, being a hub in the Expert Crowd means that a person *knows* many other experts in similar expertise areas. The main argument of our approach is that the likelihood of a successful delegation of RFSs to other experts increases based on the *hubness* of a person (embedding of a person in expert areas such as communities and interest groups).

Let us start formalizing this concept. A personalized expert query Q is defined as $Q = (KW, W(kw))$ where $KW = \{kw_1, kw_2, kw_3, \dots\}$ is the set of keywords or terms determining the context of a query. Each keyword kw may have a weight associated with it depicted by $W(kw)$. Consider the scenario in Fig. 3.7. First, a query (see Q^A and Q^B, A and B depicting the query context) is specified either manually by a (human) expert seeker or derived automatically from a given process context (PFL), for example a predefined rule denoting that a particular set of skills is needed to solve a problem. The purpose of a query is to return a set of HPSs who can process RFSs, either by working on the RFSs or delegation. Thus, Q^A would return H^A as the user who is well-connected to *authorities* in query context Q^A. Being well-connected means that H^A has the highest number of links *and* has performed interactions over these links (e.g., delegations) that are relevant for a given query context. There are two influencing factors, i.e., relations, determining hub- and authority scores: (i) how much hubs *trust* authorities (depicted as filled arrows from hubs to authorities) and (ii) *ratings* hubs receive from authorities (open arrows to hubs). Trust mainly influences the potential number of users (e.g., known by H^A) who can process delegated RFSs. On the other hand, receivers can associate ratings to RFSs to express their opinion whether the delegated RFSs fit their expertise. Q^B may demand for a different set of skills. Thus, not only matching of actors is influenced, but also the set of interactions and ratings considered for calculation expertise scores (i.e., only the set of RFSs and ratings relevant for Q^B). Note, single *interactions* that lead to trust relations, as well as single *rating actions* that lead to rating relations are not depicted by Fig. 3.7. A single arrow may in fact depict a number of interactions or ratings.

3.5.4 Ranking Model

One of the main pillars of our work is to consider the *context* in which interactions take place. In our previous work we defined two independent expertise ranking approaches, one called *DSARank* [27] and the other approach called *ExpertHITS* [28, 29]. Here we introduce a generalized ranking approach based on our previous discussions on the concept of hubs and authorities in evolving Expert Crowds. The starting point for our ranking algorithm is (3.1) (see [16, 28]).

$$H(u) = \sum_{(u,v)\in E} A(v) \qquad A(v) = \sum_{(u,v)\in E} H(u) \qquad (3.1)$$

The edge (u, v), which reads u knows v, is established based on links in the social network (FOAF profiles). Notice, by ranking nodes in a graph G using this method, each node $u \in V$ receives both hub *and* authority scores. However, we are primarily interested in computing the *hub importance* $H(u)$ of a particular node. This is motivated by the need to find coordinators who distribute requests by delegating tasks within the Expert Crowd [28] (emerging CFLs). However, we argue that an expertise mining algorithm must consider a person's interest and activity level in a certain collaboration context. As proposed in [27], *preferences* that are based on mining of interaction metrics can be used to compute contextual expertise profiles.

$$H(u;Q) = (1 - \lambda_h)p(u;Q) + \lambda_h \sum_{(u,v)\in E} w_{vu}^Q A(v;Q) \qquad (3.2)$$

Table 3.1 Interaction weights and related symbols

Symbol	Description
w_{vu}^Q	The link weight based on ratings given by v to RFSs received from u
w_{zv}^Q	The connection strength of a hub z to authority v. Delegation behavior of hubs is based on the success of interactions (successful completion of delegated task)

Computing contextual expertise profiles is accomplished by expanding (3.1) in terms of adding $(1 - \lambda_h)p(u; Q)$ to the standard HITS model as shown in (3.2). The parameter λ_h is used to balance between preferences $p(u; Q)$ and the propagation of global importance scores denoted by the term $\sum_{(u,v)\in E} w_{vu}^Q A(v; Q)$. The link weight w_{vu} based on Q is discussed in Table 3.1. From the network point of view, the definition in (3.2) can be interpreted as *influence propagation* based on a node's outgoing links. This is similar to TrustRank [12] where trust scores are propagated along neighboring outlinks. TrustRank is based on an *inverse PageRank* model that utilizes *good seeds* to influence trust flows. Also (3.2) permits a similar interpretation because $H(u; Q)$ can be computed as the inverse PageRank [12]. However, our approach closely follows the *personalized PageRank* model [21] by assigning preferences to the personalization vector $p(u; Q)$ to create context-aware importance rankings.

Similarly, importance scores for authorities $A(v; Q)$ are determined using (3.3):

$$A(v; Q) = (1 - \lambda_a)p(v; Q) + \lambda_a \sum_{(z,v)\in E} w_{zv}^Q H(z; Q) \qquad (3.3)$$

Without considering the dual nature of HITS (assigning hub and authority scores to each node in the network), we can regard (3.3) as the personalized PageRank model that is biased towards a particular interaction context using the contextual preference vector $p(v; Q)$. Again, the weight w_{zv}^Q is detailed in Table 3.1. Notice, (3.3) permits an interpretation of delegation behavior within the Expert Crowd as a stochastic process as hubs may choose to interact with known authorities or decide to pick a *newcomer* for task delegation either randomly[6] or based on, for example, interest similarities (see also [33] for bootstrapping newcomers in collaborations).

To create a unified equation for $H(u; Q)$, we substitute $A(v; Q)$—as defined in (3.3)—in (3.2) and define the hub importance of u as follows:

$$H(u; Q) = (1 - \lambda_h)p(u; Q) + \lambda_h(1 - \lambda_a) \sum_{(u,v)\in E} w_{vu}^Q p(v; Q)$$

$$+ \lambda_h \lambda_a \sum_{(u,v)\in E} \sum_{(z,v)\in E} w_{vu}^Q w_{zv}^Q H(z; Q) \qquad (3.4)$$

---

[6] The probabilistic interpretation of PageRank is known as the *random surfer* model [21].

(3.4) provides the basic formalism to determine coordinators based on contextual preferences. Next, we reformulate the context-sensitive personalization vector $p(u; Q)$ as follows (based on (3.4)):

$$p'(u; Q) = \frac{(1 - \lambda_h)}{(1 - \lambda_a)} p(u; Q) + \lambda_h \sum_{(u,v) \in E} w_{vu}^Q p(v; Q) \qquad (3.5)$$

(3.5) essentially consists of two components: preferences given to a particular hub u, for example based on the PFL problem context, and how well u is rated by authorities expressed by the weight w_{vu}^Q. The authority preference vector $p(v; Q)$ is personalized based on interaction dynamics captured by metrics such as the *interaction intensity* of v. We refer interested readers to [27] for a detailed description on these metrics.

Here we focus on personalizing $p'(u; Q)$ based on ratings to reduce the complexity of preference parameters (i.e., determining $p(u; Q)$). By setting $\lambda_h = 1$ we have:

$$p'(u; Q) = \sum_{(u,v) \in E} w_{vu}^Q p(v; Q) \qquad \text{with } \lambda_h = 1 \qquad (3.6)$$

Based on (3.4) and (3.6), let us define the following equation to estimate the hub importance of a given network node u:

$$I^H(u; T') = (1 - \lambda) p'(u; T') + \lambda \sum_{(u,v) \in E} \sum_{(z,v) \in E} w_{vu}^{T'} w_{zv}^{T'} I^H(z; T') \qquad (3.7)$$

(3.7) introduces various new concepts (detailed in Table 3.2). In particular, we define I^H as the hub importance of a node u since our approach does not require two types of rankings (hub and authority scores) anymore. Given (3.7), we have derived an expertise ranking model that is similar to the basic idea of PageRank. While such a model has been extremely successfully applied to search engines on the Web, the drawback is the complexity of computing the PageRank equation.[7] Especially in crowdsourcing scenarios that require on-demand discovery of experts based on a set of specified skills, computation of expertise scores taking up to several hours is

Table 3.2 Topic-sensitive hub importance and related symbols

Symbol	Description
I^H	The topic-sensitive hub importance score of a given node in G
T'	Topic $T' \subseteq T$ based on a set of tags applied to interactions. T' can be calculated automatically based on tag-clustering techniques (e.g., see [33]) or by using a predefined skill-based taxonomy for tags [28, 29]

---

[7] In large social networks (for example network size >10000 nodes), it may take up to several hours to compute PageRank importance scores.

not acceptable. We have first raised this issue in [26] and proposed a combination of *offline mining* and *online aggregation* of expertise ranking scores based on query preferences. Here we apply this approach to solve the problem of context-sensitive hub discovery in Expert Crowds. The first step (as shown in (3.7)) was to introduce predefined topics T' that are *query independent*.

To create topic-sensitive expertise profiles offline through mining that can be aggregated online, we propose the PageRank linearity theorem:

Theorem 1 (Linearity) *For any personalization vectors p_1, p_2 and weights w_1, w_2 with $w_1 + w_2 = 1$, the following equality holds:*

$$ppr(w_1 p_1 + w_2 p_2) = w_1 ppr(p_1) + w_2 ppr(p_2) \qquad (3.8)$$

The above equality states that personalized PageRank vectors *ppr* can be composed as the weighted sum of PageRank vectors. The linearity theorem has been originally introduced by [13, 14] to create topic-sensitive importance scores for Web-pages, but has not been applied in existing (related) approaches for expertise mining.

$$I^H(u;Q) = w_1 I^H(u;T_1) + w_2 I^H(u;T_2) \quad \text{with} \quad Q = \{T_1, T_2\} \qquad (3.9)$$

(3.9) shows how to create query-dependent rankings established upon topic-sensitive expertise importance scores using (3.7) and (3.8).

3.6 Evaluation

We structure our evaluation in four sub-sections. First, we discuss a SOA-based testbed environment allowing us to simulate crowdsourcing scenarios. Second, we present performance experiments based on logged interaction data to test the efficiency of our online ranking approach considering concurrent expertise queries. Third, we analyze the effectiveness of our ranking approach based on synthetic interaction data gathered through simulations.

3.6.1 SOA Testbed Environment

Our evaluations were gathered using the features of the Genesis2 framework [15] and infrastructure services (e.g., logging) as introduced in [24]. Genesis2 has a management interface and a controllable runtime to deploy, simulate, and evaluate SOA designs and implementations. A collection of extensible elements for these environments are available such as models of services, clients, registries, and other SOA components. Each element can be set up individually with its own behavior, and steered during execution of a test case. For the experiments in this work, we deployed

Genesis2 Backends to the *Amazon Elastic Compute Cloud*.[8] We launched, depending on the amount of involved services instances, two or three *Community AMIs* of the type *High-Memory Extra Large Instance* (17.1 GB of memory) running a Linux OS. In the following, we provided each instance with the same Genesis2 Backend snapshot via mountable volumes from the *Elastic Block Store*. Finally, we deployed the following environment setup from a local Genesis2 Frontend. It included SOA-based HPS communities established by Genesis2 Web services equipped with simulated behavior and predefined relations to provide communication channels and instantiate communities. Services act like HPSs when delegating each other new tasks, processing tasks, re-delegating existing tasks, or reporting tasks' progress status. Tasks are not delegated arbitrarily but must match the receivers capabilities. Therefore, they are tagged with three keywords one of which must match the picked receivers capabilities. Task processing and delegation decisions happen individually and in random time intervals (1–8 s). A hub combines capabilities of multiple communities by distributing tasks according to expertise areas of a given community (brokering of tasks). A hub avoids task processing and only forwards tasks. Finally, the deployed testbed environment has a variable number of services and participants per community. Consequently, the number of hubs varies depending on disparate expertise communities hubs are connected to (through knows relations).

3.6.2 Performance Aspects

We performed several experiments to test the performance of our expertise ranking algorithms under varying characteristics such as number of nodes and expertise communities. Graph-based modeling and ranking algorithms have been implemented in C# and were deployed on our local (lab-based) blade servers accessible via a query Web service.

Hardware Setup. Our servers are equipped with Intel Xeon 3.2 GHz CPUs (quad core) and 10 GB RAM hardware. Interaction logs are managed by MySQL 5.0 databases. A client request pool (RP, see Fig. 3.8a) is created on a separate machine (Intel Core2 Duo CPU 2.50 GHz, 4 GB RAM) to perform parallel invocations of the query Web service. Clients are connected with the server via a local 100 MBit Ethernet.

Performance Results. The results for *online expertise queries*[9] are summarized in Fig. 3.8. The first experiment is based on a graph containing 198 nodes, 200 edges, and a total number of 10 distinct tags applied to interactions between nodes. The query service processing time for this environment is shown in Fig. 3.8a. We vary the number of concurrent requests, denoted as RP, by launching multiple threads. Given a size of **RP = 50** and a total amount of # 100 requests to be processed, setting RP = 100 does not speed up the processing time of requests (i.e., the total time needed to process a number of requests). The average processing time increases

---

[8] http://aws.amazon.com/ec2/

[9] Performance of the offline mining procedure as discussed previously is not shown here.

Experiment	# Req.	MIN	AVG	MAX	Total
1(RP=10)	50	3167	9083	10368	52543
	100	1669	9369	10576	101244
	200	1825	9211	10748	190647
1(RP=50)	50	1606	15955	29952	50762
	100	1482	27440	48562	98685
	200	1638	36313	47689	188573
1(RP=100)	50	1606	15955	29952	50762
	100	1544	28560	57501	105331
	200	1591	55185	100370	202394
2(RP=50)	100	2308	37891	63258	123677
3(RP=50)	100	2854	42041	67516	136266
4(RP=50)	100	3276	55058	84739	167778

(a)

Applied Tag sin Exp.4(n=1029 and communities=230)	Frequ.
self-*	295
Robustness	306
Testbed	311
DB	314
Healing	321
Trust	322
WS	327
Autonomic	335
Similarity	341
Logging	353

(b)

Query ID	Query keywords	# Hubs	AVG proc.time
Q1	Robustness Logging	105	3993
Q2	Robustness Logging DB Testbed	134	3666
Q3	Robustness Logging DB Testbed Similarity	146	3478

(c)

Fig. 3.8 Processing statistics in simulated environment (in milliseconds). **a** Processing time. **b** Tag frequency. **c** Queries in Exp. 4, number of discovered hubs and AVG processing time

by comparing RP = 100 and RP = 50 due to the overhead when handling a larger amount of requests simultaneously. Thus, we use RP = 50 for all further experiments.

Also, by processing a larger amount of requests, say # 200, the total processing linearly increases with the number of requests. We increased the number of nodes and interactions to understand the scalability of the query Web service under different conditions: experiment 2 with 579 nodes, experiment 3 comprising 774 nodes, and experiment 4 with 1029 nodes in the tested. HPSs in the testbed have been deployed equally on multiple hosts, e.g., 3 cloud hosts in experiment 4 to achieve scalability. In subsequent experiments detailed in Fig. 3.8 (experiments 2–4) we focus on a request pool with RP = 50 and 100 requests to be processed by the query service using different keywords (see Fig. 3.8c). To compare the experiments 1–4, we query the interaction graph using the keywords Q = {Robustness, Logging}. Increasing the number of nodes by a factor ≈ 3 (see experiment 1 and 2), the processing time goes up by 30%. Comparing the experiments 2 and 3 (node addition of $\approx 30\%$), the processing time increases by 10%. By comparing the experiments 3 and 4 (node addition of $\approx 30\%$), the processing time increases by 20%. Our experiments show that the online creation of expertise profiles based on different queries scales with larger testbeds linearly.

Furthermore, we used different query keywords as shown in Fig. 3.8c. The number of discovered hubs increases if multiple keywords are used (see Fig. 3.8b for the set of available tags). The average processing time is not significantly influenced by the number of used keywords.

3.6.3 Quality of Expertise Rankings

Next, we analyze the effectiveness of our ranking approach based on synthetic inter-action data since real interaction logs have not been available at time when performing this research.

Ranking Evaluation Metrics. To study the results of our ranking approach, we define a set of ranking evaluation metrics in the following.

- The absolute ranking change $RC(u)$ returns the ranking change in a given query:

$$RC(u) = pos(u)_{BLR} - pos(u)_{CSR} \tag{3.10}$$

  BLR are the *base-line rankings* (here we use the standard HITS algorithm to obtain the base-line results) compared with CSR *context-sensitive rankings* using our with (cf. I^H as defined by (3.9)).
- We define *quality* $\mathscr{Q}(u)$ as the aggregated link weights of u's neighbors as:

$$\mathscr{Q}(u) = \sum_{(u,v)\in E} \sum_{(z,v)\in E} w_{zv} \tag{3.11}$$

  We have studied the calculation of link weights extensively in our previous work. For example, weights can be calculated based on trust metrics (e.g., delegation behavior) or link intensity [27]. Thus, we refer the interested reader to [28, 33].

Algorithm Parameters. CSR are obtained based on both link weights and the assignment of preferences to personalization vectors $p'(u;T') = \sum_{(u,v)\in E} w_{vu}^{T'}$ $p(v;T')$. In our experiments, preferences are assigned as follows:

$$p(v;T') = \begin{cases} 1, & \text{if } T'[v] \neq \texttt{null} \\ 0, & \text{otherwise} \end{cases} \tag{3.12}$$

$T'[v]$ holds those users who have interacted with other users with focus on a particular topic T'. For example, users have performed tasks tagged with keywords related to T'. However, not only interaction-based profiles must be used to assign preferences. In addition, a user's manually maintained profile (e.g., FOAF) may be used to account for the user's interest (i.e., the authority v) in a given topic.

Ranking Results. To test the effectiveness of CSR, we performed experiments to study the impact of ratings and link weights on expert rankings. In the following figures, we show the top-30 ranked experts in a small-scale network (100 nodes). Results are sorted based on the *position* within the result set (see horizontal axis of Fig. 3.9 and column **Rank** in Fig. 3.10). Figure 3.9a shows the node degree and

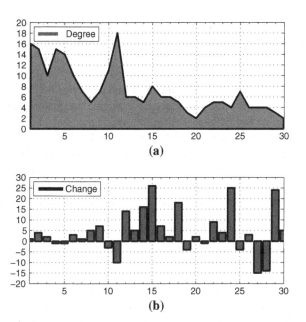

Fig. 3.9 Node degree and results of CSR/BLR comparison. **a** Node degree. **b** Ranking change

Rank	Quality $\mathcal{Q}$	Rating	Rank	Quality $\mathcal{Q}$	Rating
1	3.7	0.8	16	1.0	1.3
2	3.0	0.7	17	1.0	0.9
3	3.0	1.4	18	1.0	0.9
4	3.0	0.5	19	1.0	2.7
5	3.0	0.5	20	1.0	4.1
6	1.0	0.8	21	1.0	1.5
7	1.0	1.6	22	1.0	1.0
8	1.0	1.8	23	1.0	0.9
9	1.0	0.8	24	1.0	1.1
10	1.0	0.3	25	1.0	0.2
11	0.4	0.9	26	1.0	1.3
12	1.0	1.1	27	1.0	1.2
13	1.0	0.9	28	1.0	0.8
14	1.0	1.1	29	1.0	1.5
15	1.0	0.3	30	1.0	2.5
(a)			**(b)**		

Fig. 3.10 CSR ranking results: rank, quality, and ratings. **a** Hub quality and ratings (1–15). **b** Hub quality and ratings (16–30)

Fig. 3.9b ranking changes obtained by comparing CSR results with BLR (i.e., ranking results without accounting for metrics and ratings).

Figure 3.10 shows that all nodes within the top segment received high ratings given a high degree of links which is the desired property of CSR. Different levels of quality (i.e., quality mainly being 1 of ranked nodes between the positions 6–30) can be explained by the impact of node degree on quality. Some nodes are demoted (negative ranking change) since the node (e.g., see 11) has received low ratings even though the node has a high degree of links. Nodes get promoted (positive ranking change) if they exhibit sufficient high ratings (see 15) or high quality (see 20 which was promoted a few positions only due to limited degree). Overall, CSR exhibit the demanded properties of promoting well-connected and rated hubs, thereby guaranteeing the discovery of reliable entry points to the Expert Crowd.

3.7 Conclusion and Future Work

The Web is evolving rapidly by allowing people to publish information and services. At the heart of this trend, interactions become increasingly complex and dynamic spanning both humans and software services. However, the transformation of how people collaborate and interact on the Web has been poorly leveraged in existing service-oriented architectures. The benefit of the presented approach is a seamless service-oriented infrastructure of human- and software services. The resulting service-oriented application needs to be flexible supporting adaptive interactions.

In this chapter, we have motivated the need for adaptive interactions discussing an Expert Crowd scenario where people can register their skills and capabilities as services. Mixed service-oriented systems are open ecosystems comprising human- and software-based services. We discussed the HPS architecture enabling dynamic interactions in mixed service-oriented systems. We defined a novel expertise ranking approach that is based on context-aware interactions. Our ranking approach shows promising results, but needs to be further validated in real crowdsourcing environments. Our future work includes the public deployment and evaluation of the implemented framework. Also, we will further study the effectiveness and quality our expertise ranking approach in large-scale collaboration environments.

References

1. Adams, M., Hofstede, A.H.M., Edmond, D., Aalst, W.M.P.V.D.: Worklets: A service-oriented implementation of dynamic flexibility in workflows. In: OTM Conferences, vol. 1, 291–308 (2006)
2. Agichtein, E., Castillo, C., Donato, D., Gionis, A., Mishne, G.: Finding high-quality content in social media. In: WSDM, pp. 183–194. ACM, New York (2008)
3. Agrawal, A., et al.: WS-BPEL Extension for People (BPEL4People), V1.0 (2007)
4. Amend, M., et al.: Web Services Human Task (WS-HumanTask), Version 1.0 (2007)

5. Balthazard, P.A., Potter, R.E., Warren, J.: Expertise, extraversion and group interaction styles as performance indicators in virtual teams: how do perceptions of it's performance get formed? Database **35**(1), 41–64 (2004)

6. Breslin, J., Passant, A., Decker, S.: Social web applications in enterprise. Soc. Semant. Web **48**, 251–267 (2009)

7. Cugola, G., Nitto, E.D., Fuggetta, A., Ghezzi, C.: A framework for formalizing inconsistencies and deviations in human-centered systems. ACM Trans. Softw. Eng. Methodol. **5**(3), 191–230 (1996)

8. Dom, B., Eiron, I., Cozzi, A., Zhang, Y.: Graph-based ranking algorithms for e-mail expertise analysis. In: DMKD, pp. 42–48. ACM, New York (2003)

9. Easley, D., Kleinberg, J.: Networks, Crowds, and Markets: reasoning About a Highly Connected World. Cambridge University Press, Cambridge (2010)

10. Garlan, D., Poladian, V., Schmerl, B.R., Sousa, J.P.: Task-based self-adaptation. In: WOSS, pp. 54–57. New York (2004)

11. Gentry, C., Ramzan, Z., Stubblebine, S.: Secure distributed human computation. In EC'05, pp. 155–164. ACM, New York (2005)

12. Gyöngyi, Z., Molina, H.G., Pedersen, J.: Combating web spam with trustrank. In: VLDB, pp. 576–587. ACM, Toronto (2004)

13. Haveliwala, T.H.: Topic-sensitive pagerank. In: WWW, pp. 517–526. ACM, New York (2002)

14. Jeh, G., Widom, J.: Scaling personalized web search. In WWW, pp. 271–279. ACM, New York (2003)

15. Juszczyk, L., Dustdar, S.: Script-based generation of dynamic testbeds for soa. In: ICWS '10, IEEE Miami (2010)

16. Kleinberg, J.M.: Authoritative sources in a hyperlinked environment. J. ACM **46**(5), 604–632 (1999)

17. Kosorukoff, A., Goldberg, D.E.: Genetic algorithms for social innovation and creativity. Technical report, University of Illinois at Urbana-Champaign (2001)

18. Kumar, A., Aalst, W.M.P.V.D., Verbeek, E.: Dynamic work distribution in workflow management systems: How to balance quality and performance. J. Manage. Inf. Syst. **18**(3), 157–193 (2002)

19. Mendling, J., Ploesser, K., Strembeck, M.: Specifying separation of duty constraints in bpel4people processes. In: Business Information Systems, pp. 273–284. LNBIP. Springer, Berlin (2008)

20. Moody, P., Gruen, D., Muller, M.J., Tang, J., Moran, T.P.: Business activity patterns: a new model for collaborative business applications. IBM Syst. J. **45**(4), 683–694 (2006)

21. Page, L., Brin, S., Motwani, R., Winograd, T.: The PageRank citation ranking: bringing order to the Web. Technical report, Stanford Digital Library Technologies Project (1998)

22. Panteli, N., Davison, R.: The role of subgroups in the communication patterns of global virtual teams. IEEE Trans. Prof. Commun. **48**(2), 191–200 (2005)

23. Petrie, C.: Plenty of room outside the firm. Internet Comput. **14**, 92–96 (2010)

24. Psaier, H., Juszczyk, L., Skopik, F., Schall, D., Dustdar, S.: Runtime behavior monitoring and self-adaptation in service-oriented systems. In: SASO, IEEE, Pisa (2010)

25. Russell, N., Aalst. W.M.P.V.D.: Evaluation of the bpel4people and ws-humantask extensions to ws-bpel 2.0 using the workflow resource patterns. Technical report, BPM Center Brisbane/ Eindhoven (2007)

26. Schall, D.: Human interactions in mixed systems—architecture, protocols, and algorithms. Ph.D. thesis, Vienna University of Technology (2009)

27. Schall, D.: Expertise ranking using activity and contextual link measures. Data Knowl. Eng. **71**(1), 92–113 (2012). doi:10.1016/j.datak.2011.08.001

28. Schall, D., Skopik, F.: Mining and composition of emergent collectives in mixed service-oriented systems. In: CEC '10, IEEE, Shanghai (2010)

29. Schall, D., Skopik, F., Dustdar, S.: Expert discovery and interactions in mixed service-oriented systems. IEEE Trans. Serv. Comput. **5**(2), 233–245 (2012). doi:10.1109/TSC.2011.2

30. Schall, D., Truong, H.-L., Dustdar, S.: Unifying human and software services in web-scale collaborations. IEEE Internet Comput. **12**(3), 62–68 (2008)
31. Shahaf, D., Horvitz, E.: Generalized task markets for human and machine computation (2010)
32. Shetty, J., Adibi, J.: Discovering important nodes through graph entropy the case of enron email database. In: LinkKDD, pp. 74–81. ACM, New York (2005)
33. Skopik, F., Schall, D., Dustdar, S.: Modeling and mining of dynamic trust in complex service-oriented systems. Inf. Syst. **35**, 735–757 (2010)
34. Skopik, F., Schall, D., Dustdar, S.: Trustworthy interaction balancing in mixed service-oriented systems. In: SAC '10, pp. 799–806. ACM, New York (2010)
35. Su, Q., Pavlov, D., Chow, J.-H., Baker, W.C.: Internet-scale collection of human-reviewed data. In: WWW '07, pp. 231–240. ACM, New York (2007)
36. Thomas, J., Paci, F., Bertino, E., Eugster, P.: User tasks and access control over web services. In: ICWS'07, pp. 60–69. IEEE, Salt Lake City (2007)
37. von Ahn, L.: Games with a purpose. IEEE. Computer **39**(6), 92–94 (2006)
38. Yang, J., Adamic, L., Ackerman, M.: Competing to share expertise: the taskcn knowledge sharing community. In: International Conference on Weblogs and Social, Media (2008)
39. Zhang, J., Ackerman, M.S., Adamic, L.: Expertise networks in online communities: structure and algorithms. In: WWW, pp. 221–230. ACM, New York (2007)

Chapter 4
Crowdsourcing Tasks in BPEL4People

Abstract In this chapter we extend BPEL4People with non-functional properties that allow to cope with the inherent dynamics of crowdsourcing processes. Such properties include human capabilities and the level of skills. We discuss the formation of social networks that are particularly beneficial for processing extended BPEL4People tasks. Furthermore, we present novel approaches for the automated assignment of tasks to a social group. The feasibility of our approach is shown through a proof of concept implementation of various concepts as well as simulations and experiments to evaluate our ranking and selection approach.

Keywords Crowdsourcing · BPEL4People · Non-functional properties · Social networks

4.1 Introduction

Most efforts to model human interactions using BPEL4People [3] (B4P) and WS-HumanTask [4] (WS-HT) focus on relatively static role models for selecting the right person to interact with. Thus, BPEL4People is not well suited for specifying and executing processes involving crowdsourcing of tasks to online communities. The WS-HT specification does not define any particular mechanisms to find or select people in open and dynamic environments. Instead, a *Logical People Group* is used to query an organizational directory. We believe that human interactions in SOA need to be supported in a flexible manner, in particular, it should be possible to use crowdsourcing for process execution.

In this chapter we present the following key contributions:

1. *An approach* for combining crowdsourcing techniques and B4P related XML standards.
2. *B4P with non-functional properties* for adaptive and quality-aware crowdsourcing of service-oriented processes.

D. Schall, *Service-Oriented Crowdsourcing*, SpringerBriefs in Computer Science, DOI: 10.1007/978-1-4614-5956-9_4, © The Author(s) 2012

3. *Social community formation* for efficient crowdsourcing of processes.
4. *Automated matching* of tasks to members of a social crowdsourcing community.
5. *Implementation and evaluation* of our concepts.

This chapter is structured as follows: In Sect. 4.2 we discuss related work, in Sect. 4.3 we present the application areas of our proposed service-oriented crowdsourcing environment and outline the steps of our approach. In Sect. 4.4 we propose extensions to introduce non-functional requirements for B4P. In Sect. 4.5 we explain the establishment of a social network structure that allows for an efficient execution of B4P processes. We present the automated matching of extended B4P processes to a social community crowd in Sect. 4.6. Finally, we present in Sect. 4.7 the results of our evaluation and conclude this chapter in Sect. 4.8.

4.2 Background

The work presented in this chapter focuses on a methodology and tools allowing human interactions in SOA to be executed in a flexible manner. There has been a growing interest in the complex 'connectedness' of today's society. Phenomena in our online-society involve networks, incentives, and the aggregate behavior of groups [15]. Peer-production distinguishes from the property- and contract-based models of firms and markets. Research by [7] analyzes how groups of individuals successfully collaborate on large-scale projects following a diverse cluster of motivational drives and social signals. *Human computation* is motivated by the need to outsource certain steps in a computational process to humans [16, 39]. A variant of human computation called 'games that matter' was introduced by [48]. Crowdsourcing [8, 13, 18, 49] refers to a new Web-based collaboration model where human tasks are outsourced to an anonymous workforce by various requesters including companies or individuals. Related to crowdsourcing are systems such as Amazon Mechanical Turk[1] (AMT). AMT is a Web-based, task-centric platform in which users can publish, claim, and process tasks. [46] evaluates the task properties of a similar platform in cases where large amounts of data are reviewed by humans. In contrast to common question/answer (Q/A) forums, such as Yahoo! Answers,[2] AMT enables businesses to access the manpower of thousands of people using a Web services API. Mixed service-oriented systems [38, 39] target flexible interactions and compositions of *Human-Provided Services (HPS)* and *Software-Based Services (SBS)* [42]. This approach is aligned with the vision of the Web 2.0, where people can actively contribute services. In such networks, humans may participate and provide services in a uniform way by using the HPS framework [39]. In a similar spirit, [33] defines *emergent collectives* which are networks of interlinked valued nodes (services). In such collectives, there is an easy way to add nodes by distributed actors so that the network will scale.

---

[1] http://www.mturk.com/

[2] http://answers.yahoo.com/

Current crowdsourcing platforms offer very limited support for modeling complex interactions that require coordination of humans' joint capabilities and software-based services.

Several trends originated from human interactions in service-oriented systems. As mentioned before, B4P defines human interactions in business processes via the human task specification [4]. A concrete implementation of B4P as a service has been introduced in [47], but without supporting process adaptivity. Worklets [1] grounded in *activity theory* represent self-contained subprocesses. Another approach for flexible activities in business-oriented environments was presented in [11]. In [36], the relation of various B4P-related Web standards and resource patterns is discussed. In contrast to process-centric compositions in SOA, task-based crowdsourcing platform such as AMT do not support long-running interactions and compositions of humans and services. The problem of composition is strongly related to organization and control. The key principles of autonomic computing [19] aim at supporting systems featuring *self-** properties such resilience through self-organizing computing elements. As an example, [26] focus on autonomic services and trusted service selection. In [25], a reference architecture for self-organizing service-oriented systems is presented, but without considering humans 'as part' of the system. The authors in [17] propose adaptive flows to support flexibility and evolution in collaborative, pervasive environments.

Human tasks metrics in workflow management system have a long history in research (e.g., see [12, 23, 52]). Studies on distributed teams focus on human performance and interactions [5, 32]. With the emergence of Enterprise 2.0 environments [10], the information available in social networks becomes important in a professional context as well. Thus, both *technological* and *social aspects* shape the operation constraints of a system [22]. As a consequence, it is important not only to model human interactions in process-centric systems, but also to understand how people are connected [43, 50] and how information flows are influenced by social structure. When building Web-centric applications involving human tasks, engineers have to consider incentive schemes that are likely to encourage users to perform these tasks that crucially rely on human input [44]. Models and algorithms to track people's expertise are important in future service-oriented crowdsourcing environments [40]. In Web-based environments, task-based platforms allow users to share their expertise [50] or help other users in Q/A communities [2]. In [51], the authors applied PageRank [31] in online communities to measure expertise.

In our previous work, we have designed and implemented the HPS framework [42] allowing users to define services and to provision human expertise in a service-oriented manner. Furthermore, we have designed a market-based crowdsourcing platform [37] and simulation environment to stimulate the evolution of user skills. Here we extend B4P and related XML standards to cope with the inherent dynamics of crowdsourcing environments. Also, we show how social networks can help to process crowdsourced tasks in a more efficient manner. To our best knowledge, this is the first effort to extend B4P and related standards to cope with the challenges of crowdsourcing environments.

4.3 Service-Oriented Crowdsourcing

Business processes are affected by rapidly changing requirements and imperative adaptations that come along with necessary modernizations of the in-house activities and adjustments to the market. Many of today's workflow-based systems are still based on a *top-down* design for processes. It is clear, that there is a trend to the combination of interactions between humans and software based applications, such as SBS, as a central requirement in business environments. This may work fairly well for processes involving only SBS with minor human interaction. However, once the human interaction models in those processes become more important and complex, a top-down approach is unable to foresee and cope with the implications of the human behavior related dynamics. There are several types of tasks that are still best processed by humans.

4.3.1 Task-Based Crowdsourcing Markets

Currently, the extensions to the specifications for business processes are designed for simpler human tasks, e.g., making process progress decisions and process approval requests [3]. Nevertheless, with the new marketplaces provided by crowdsourcing including workers with manifold skills, new types of tasks can be considered for outsourcing. Recently, many platforms have started to offer a versatile number of tasks that can be outsourced to the crowd.

In the following, we overview some of the potential crowd tasks that could be designed and outsourced using our approach:

- *Classification* or *categorization* tasks using for example the AMT marketplace [20], CrowdFlower,[3] or SmartSheet.[4] Categorization is one of the most common use cases for crowdsourcing. A categorization task is one that asks a worker to select from a list of options.
- *Transcription* tasks as offered by CastingWords[5] or SpeechInk.[6] Transcription services include transcription of audio-to-text.
- *Web development* and web programming as provided by, for example, oDesk.[7] Web development tasks include integration of scripts with Web service APIs or programming questions regarding different frameworks or toolkits.

We foresee that our B4P-based approach can be used in scenarios like document translation, proof-reading and correction of documents (see also crowd-powered

---

[3] http://www.crowdflower.com

[4] http://www.smartsheet.com

[5] http://www.castingwords.com/

[6] http://www.speechink.com/

[7] http://www.odesk.com

word processors[8]), transcription, data-cleansing, and simple programming tasks as those offered by oDesk. These tasks can be crowdsourced by creating appropriate B4P activities and WS-Human Tasks embeddings that are transmitted to the HPS middleware. The HPS middleware implements algorithms for matching, ranking, selection and runtime monitoring of tasks.

4.3.2 Approach Outline

Here, we propose adaptive human interaction support in service-oriented systems. Human interaction support in SOA has only recently been proposed and supported by standards such as WS-HT [4] and B4P [3]. We argue that these standards need to be extended to support compositions of both HPS and SBS in crowdsourcing applications. The benefit of this approach is a seamless QoS-based service-oriented infrastructure that is able to adjust its interactions based on service-level agreements (SLAs) and quality constraints. Our previous work [37, 39] has already detailed the challenges of integrating in-house processes to a crowdsourcing environment. Here the approach is extended by support for crowdsourcing of composed, complex tasks. Here we identify the following main challenges:

- *Crowd structure*: Composed tasks not only require humans for processing the set of subtasks, but also, a coordinated and supervised assignment and merging of the individual results to a final result. For this purpose, in this work we identify three roles. The *Coordinator*, on the one side, keeps in touch with the business process management, and, on the other side, maintains her/his relations to various crowd communities. The motivation of coordinators is to some extend similar to the role of a moderator in social-online communities such as *Slashdot*. Moderators, in our case coordinators, can be understood as gatekeepers [24] who control the quality of postings (in our case tasks) in online communities. The *Supervisor* represents a community and is aware of the current possible segmentation of one crowd task to a set of subtasks with the related distribution to her/his team. Finally, there are the common crowd *Workers*. We believe that this simple role model consisting of three distinct roles is sufficient for crowd tasks as described in Sect. 4.3.1. Crowd members temporarily form teams to work jointly on tasks on a magnitude of hours and days. Shortly after that the team dissolves again and workers pick some other task. More complicated hierarchies as found in enterprises or large-scale development teams may not be suitable for such short-lived groups.
- *Non-functional properties*: Current service-based process definitions for human interactions, in particular the combination of BPEL, B4P, and WS-HT definitions, need to be extended for the requirements of crowdsourcing. A further challenge addressed with crowdsourcing, is the integration of SOA agreements' specifications WSLA (Web Service Level Agreements)[9] in the outsourcing process.

---

[8] http://projects.csail.mit.edu/soylent/

[9] http://www.research.ibm.com/wsla/WSLA093.xsd

- *Crowd member ranking*: Crowd environments are dynamic by nature. Therefore, it is vital to the outsourcing party that the current best matching crowd members can be detected and ranked according to the task's requirements. The result allows the customer to select from a large set of potential workers. Also, the final decision remains with the customer that can hide possibly sensitive selection constraints from the public crowd platform.

An important aspect of this work is to introduce the notion of *expertise* (i.e., human skills) in the context of B4P. Existing approaches for expertise mining (e.g., see [51]) have mostly been applied in online communities or social network analysis, but not in process-oriented crowdsourcing environments. Here we introduce the notion of *capabilities* and *skills* in B4P to ensure quality-aware crowdsourcing of human tasks. In the following, we define important concepts used in this work:

- *Skills* are specific to the functions workers perform doing their job. As an example, a worker may perform activities related to a software development task such as reviewing code. The worker may be an expert in 'Java programming', a beginner in 'Python programming' and so forth. However, skills—as used in this work—are always based on *personal expertise* workers have and workers may improve their skills through training (e.g., improving the skill level from 'knowledgeable' to 'expert').
- *Capabilities* describe non-functional human properties to determine a workers suitability to work on a task. Human capabilities describe behavior properties which cannot be directly derived from the worker's profile. Example capabilities include 'worker should be capable of coordinating 5 other team members' or 'worker should be capable of merging and finalizing translated input from other workers'. Thus, the suitability of a worker is highly dependent on the current load conditions within the crowdsourcing environment. A worker might have in principle the skills to finalize a translated document but may not have the resources to merge the input from other workers. At B4P level, only capabilities are defined such as 'translate document and split/merge sections'. To fulfill this capability, workers are matched and ranked based on their skills, their social network connectivity (being able to split, distribute portions of the document to peers in the social network and to merge the received input) and their current load (number of pending tasks in a worker's queue).
- *Constraints* allow the customer and owner of the WS-HT to state some strict or relaxed filter options on the different roles. For example, from the customer point of view, certain crowd members may belong to a group *Preferred Workers* that is either populated automatically based on past experiences (e.g., reliable and trustworthy behavior of a worker towards the customer) or other customer internal policies. Constraints could state that only *Preferred Workers* should be able to review certain parts of a document. Other constrains could for example state that only workers from a certain geographical location may work on a task.

Figure 4.1 outlines the idea of the approach. In the first step in Fig. 4.1a, part of a BPEL process includes a B4P extended activity (ba4) to transfer a set of human tasks

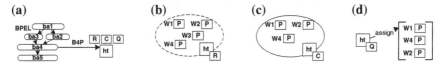

Fig. 4.1 Enhanced B4P environment: matching, ranking, and selection of human workers. **a** BPEL/B4P processes. **b** FP matching. **c** NFP matching. **d** Select and assign

(ht) to the crowd. A task's description comprises functional properties (FPs), e.g., assignment regions R, and furthermore, non-functional properties (NFPs) including capabilities C and quality expectations Q. In Fig. 4.1b a set of potential crowd workers W that can participate in the task ht is estimated by matching the task's assigned set of regions to the regions available in the environment. Next, in Fig. 4.1c the initial set of workers is reduced to a set that provides the required NFPs, e.g., their capabilities. Additionally, in this step the workers are ranked according to their capabilities' related skill level. The skill level hints the expectable quality of returned task results. It is important to note that the requester has no knowledge about the hierarchical structure in the crowd. Hence, the workers recommended to the requester are actually a set of coordinators with aggregated capabilities. Finally, to guarantee the promised quality the human tasks are assigned according to the ranking (see Fig. 4.1d).

All interactions within the environment are monitored by a logging component. These logs contain information regarding task creation time, task assignments, interactions among workers and so forth. Based on this information various metrics are automatically calculated such as task processing speed and reliability (accepted and finished tasks). The ranking procedure is based on these metrics which are frequently updated. Thus, this feedback-loop approach enables the system to self-adapt based on the workers' behavior.

4.4 Non-Functional Properties in B4P

Current B4P compositions include mainly functional properties. In the common case these comprise the WSDL based information (operations and ports), and, related to the potential assignments, role-based access models to the activities around the task (c.f. [3, 27, 47]). However, the dynamic nature of crowdsourcing environments requires a flexible definition of interactions in B4P. In particular, a situation-aware selection of potential workers must be possible. Thus, instead of defining strict interaction models it is necessary to include in the definitions some properties that guarantee a certain degree of freedom at composition and execution time. We call these particular properties *non-functional properties* for B4P. Just as the functional properties, these properties define possible task assignments and human task processors that come into consideration. However, non-functional properties' values are either not completely known a-priori, or tend to change rather frequently over time. A crowd worker's observed performance might be better or

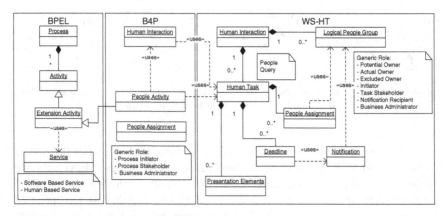

Fig. 4.2 Simplified B4P task model

less than expected because of the worker's current situation and context which influences her/his behavior. A typical example includes her/his present task load. The following section gives a brief overview of the concepts included in BPEL, B4P, and WS-HT before the new extensions are presented and explained in an XML example.

4.4.1 Human Tasks in B4P

Figure 4.2 shows the relation between BPEL, B4P, and WS-HT. The figure is a simplified version of the relation and contains only the essential elements that are addressed by our extensions.

Generally, a BPEL *Process* uses services to process the included activities. In our case these can either include SBSs or HPSs. In the event of an HPS activity, the B4P specification allows to bridge a BPEL *Extension Activity* to a B4P *People Activity*. The *People Activity* contains either locally defined tasks from B4P *Human Interaction*, a container for task definitions, or, from a WS-HT document's *Human Task*. Apart from the wrapping tasks, B4P also defines *People Assignments*. These are defined role types that refer to the whole process context. Similar to B4P's definition, WS-HT has a *Human Interaction* element. It is also a container for a task collection, however, with additional elements. In particular, it allows to define *Logical People Groups* that list the involved parties and their roles aligned with the tasks. The *Human Task* used by both B4P and WS-HT is actually a WS-HT element that defines the individual assignments between task and human (*People Assignment*), deadlines with the *Deadline* element, and human readable description of the task with the *Presentation Elements*.

4.4.2 Basic Model and Extensions

The current specification of WS-HT provides no elements to include non-functional properties into their definition as required in crowdsourcing. This is a major short-coming in the specification when applied to dynamic environments. The reminder of this section explains the extensions to WS-HT necessary to include non-functional properties. Additionally, in reference to the previous work in [34] with a similar scenario, we consider that the agreements between the crowd brokers, the previously introduced *Coordinator*s, and the customers settle on an XPath processable WSLA document. Hence, the values to the WS-HT functional and non-functional properties are manly taken from an WSLA XML document.

```
1   <?xml version="1.0" encoding="UTF-8"?>
2   <htd:humanInteractions xmlns:htd="http://www.example.org/WS-HT"
3     xmlns:xsi="http://www.w3.org/2001/XMLSchema-instance"
4     xmlns:xsd="http://www.w3.org/2001/XMLSchema" xmlns:tns="http://danielschall.at/hps/"
5     targetNamespace="http://danielschall.at/hps/" xmlns:cp="http://danielschall.at/crowd/params"
6     targetNamespace="http://danielschall.at/crowd/params"
7     xsi:schemaLocation="http://danielschall.at/WS-HT/ws-humantask.xsd">
8     <htd:import importType="http://schemas.xmlsoap.org/wsdl/"
9       location="CrowdSourcing.wsdl" namespace="http://danielschall.at/crowd" />
10    <!-- htd:logicalPeopleGroups -->
11    <!-- htd:tasks -->
12    <htd:notifications>
13      <htd:notification name="taskComplete">
14        <htd:interface portType="tns:CustomerPT"
15        operation="reportComplete" />
16        <htd:peopleAssignments>
17          <htd:recipients>
18            <htd:from>htd:getInput("SLA")//wsla:Parties/wsla:ServiceConsumer</htd:from>
19          </htd:recipients>
20        </htd:peopleAssignments>
21        <htd:presentationElements/>
22      </htd:notification>
23    </htd:notifications>
24  </htd:humanInteractions>
```

Listing 4.1 Human interactions including roles, tasks and notifications

Human Interaction. Listing 4.1 presents an extract of a WS-HT XML docu-ment. Such a document starts with the *humanInteractions* tag and for our purpose links an additional namespace (*cp*). Next, the namespace (*tns*) related to the crowd-sourcingservice is defined. Then, the *import* tag specifies the WSDL file and its location for all WS operations specified in the interaction. The following *logicalPeo-pleGroups* and *tasks* will be detailed by the following listings. At the end, an example of a notification is listed. In WS-HT notifications are used to notify a person or a group of people of a noteworthy business event. For this crowdsourcing scenario a general one is defined at the end of the interaction definition. Via the port *CustomerPT* the customer is informed that her/his outsourced task is complete. In this case, the *from* tag does not specify one of the logical people groups defined in Listing 4.2. Instead, it specifies the service customer which can be found in the WSLA's *Parties* section.
Logical People Groups. This part of the human interactions definition organizes the members of the crowd in our scenario in groups of people. As motivated in

Sect. 4.3.2 and in detail explained later in Sect. 4.5 we have identified three distinct roles for structured crowdsourcing. Listing 4.2 defines those roles for the WS-HT document. Furthermore, the content of the tag *logicalPeopleGroup* allows to characterize the different roles in the outsourcing process. Here, our first extension to the standard is evident. Apart from the standard WS-HT parameter `region`, the parameters define some necessary non-functional properties for the groups.

```
 1  <htd:logicalPeopleGroups>
 2    <htd:logicalPeopleGroup name="taskCoordiantors">
 3      <htd:documentation xml:lang="en-US">
 4        coordinate tasks in the crowd
 5      </htd:documentation>
 6      <htd:parameter name="region" type="xsd:string" />
 7      <htd:parameter name="communities" type="cp:ListOfCommunities" />
 8      <htd:parameter name="capabilities" type="cp:tListOfAggregateCapabilities" />
 9      <htd:parameter name="constraints" type="cp:tListOfConstraints" />
10    </htd:logicalPeopleGroup>
11    <htd:logicalPeopleGroup name="taskSupervisors">
12      <htd:documentation xml:lang="en-US">
13        supervises tasks in the crowd and aggregates results
14      </htd:documentation>
15      <htd:parameter name="region" type="xsd:string" />
16      <htd:parameter name="communityId" type="xsd:string" />
17      <htd:parameter name="capabilities" type="cp:tListOfAggregateCapabilities" />
18      <htd:parameter name="constraints" type="cp:tListOfConstraints" />
19    </htd:logicalPeopleGroup>
20    <htd:logicalPeopleGroup name="taskWorkers">
21      <htd:documentation xml:lang="en-US">
22        can processes tasks
23      </htd:documentation>
24      <htd:parameter name="region" type="xsd:string" />
25      <htd:parameter name="communityId" type="xsd:string" />
26      <htd:parameter name="capabilities" type="cp:tListOfCapabilities" />
27      <htd:parameter name="constraints" type="cp:tListOfConstraints" />
28    </htd:logicalPeopleGroup>
29  </htd:logicalPeopleGroups>
```

Listing 4.2 Groups defined in the context of the crowd model

These include a parameter for the affiliation of the crowd members. For the coordinator this is a list of `communities`. The `communityId` is for members affiliated with exactly one community (supervisor and worker). Next, the capabilities of a member are stated in `capabilities`. Those of the coordinator and supervisor are aggregated across the social hierarchy. Finally, a property denoted `constraints` allows the customer and owner of the WS-HT to state some strict or relaxed filter options on the different roles. Hard constraints may be used to hide sensitive information from crowd members and soft constraints may be used to influence the automatic matching of tasks to crowd members.

Tasks. The main part of a WS-HT states the tasks and subtasks. Listing 4.3 includes a single document review task, a service offered by many current crowdsourcing platforms. After a brief documentation of the content, the WS port offering the review service is defined. The port connects the caller to the selected crowd coordinator. Any further delegation of the task is not directly influenced by this task specification, but remains in the hands of the crowd members. Another necessary extension to the standard is the *quality* tag. It defines the expected quality of the promised reviews. The value of the quality is derived from the publicly accessible WSLA document with immutable content and prepared to be reused for future similar review tasks. Therefore, there is a need for a document that contains

```
1   <htd:tasks>
2     <htd:task name="DocumentReview">
3       <htd:documentation xml:lang="en-US">
4         This task requires the review of a multipage document
5       </htd:documentation>
6       <htd:interface portType="tns:ReviewHandlerPT" operation="review"
7         responsePortType="tns:ReviewHandlerCallbackPT" responseOperation="reviewResponse" />
8       <cp:quality>
9       htd:getInput("SLA")//wsla:Expression[wsla:SLAParameter='quality']/wsla:value
10      </cp:quality>
11      <htd:peopleAssignments>
12        <htd:taskStakeholder>
13          <htd:from logicalPeopleGroup="taskCoordiantors">
14            <htd:argument name="region"> <!-- hard constraints -->
15              htd:getInput("SLA")//wsla:Expression[wsla:SLAParameter='region']/wsla:value
16            </htd:argument>
17            <htd:argument name="capabilities"> <!-- hard constraints -->
18              htd:getInput("SLA")//wsla:Expression[wsla:SLAParameter='capabilities']/wsla:value
19            </htd:argument>
20            <htd:argument name="constraints"> <!-- hidden hard/soft constraints -->
21              htd:getInput("DocumentReview")/listOfConstraints
22            </htd:argument>
23          </htd:from>
24        </htd:taskStakeholder>
25      </htd:peopleAssignments>
26      <htd:presentationElements>
27        <htd:name xml:lang="en-US">Document Review</htd:name>
28        <htd:presentationParameters>
29          <htd:presentationParameter name="title" type="xsd:string">
30          htd:getInput("DocumentReview")/title
31          </htd:presentationParameter>
32          <htd:presentationParameter name="chapters" type="xsd:integer">
33          htd:getInput("DocumentReview")/chapters
34          </htd:presentationParameter>
35          <htd:presentationParameter name="document_url" type="xsd:string">
36          htd:getInput("DocumentReview")/document_url
37          </htd:presentationParameter>
38          <htd:presentationParameter name="review_url" type="xsd:string">
39          htd:getInput("DocumentReview")/review_url
40          </htd:presentationParameter>
41          <htd:presentationParameter name="result_url" type="xsd:string">
42          htd:getInput("DocumentReview")/result_url
43          </htd:presentationParameter>
44          <!-- and more -->
45        </htd:presentationParameters>
46        <htd:subject xml:lang="en-US">
47          Review of a document {$title} comprising {$chapters}
48        </htd:subject>
49        <htd:description xml:lang="en-US" contentType="text/plain">
50          Review the attached document {$title} comprising {$chapters}.
51          Find the document at {$document_url} and the related questionnaire at {$review_url}.
52          Only fully and in-time completed questionnaires accepted at {$result_url}.
53        </htd:description>
54      </htd:presentationElements>
55      <!-- htd:deadlines -->
56      <!-- crowdsourcing is flexible thus NO task compositions -->
57    </htd:task>
58  </htd:tasks>
```

Listing 4.3 The set of human tasks

the customers requirements for this particular task. In the presented case, this is the
`DocumentReview` document only accessible by the customer.

In the tag *peopleAssignments*, after region and capabilities criteria, the tag
`constraints` contains a number of hard or soft constrains in *listOfConstraints*
that define the final selection and are provided by `DocumentReview`. These might
include constraints that must not be propagated to the crowd. Otherwise, as by definition the content of the tag *peopleAssignments* maps between the roles defined in
WS-HT and their properties defined in Listing 4.2.

The *presentationElements* tag contains standard information about human tasks
and notifications. This is another example where all values are gathered from the
customers own document. As intended by the standard, the content is human-readable
information about the task and structured according to content displayable at a user
interface. This allows the customer to specify the particularities of the task and the
involved crowd members to deal with their tasks and notifications via a user interface.

From Listing 4.3 it becomes apparent that the document comprises a number of
chapters. Also, urls are defined indicating where the document is provided for review
(`document_url`), the location of the questionnaire (`review_url`), and a result
submission portal at `result_url`. According to the WS-HT standard, the *task*
tag can also include compositions with a set of subtasks connected to individual
task definitions. These tags provide a convenient method to detail segmented tasks
a-priori. Nevertheless, crowdsourcing is a dynamic environment and a top-down
segmentation of the task might contradict the current possible worker assignments.
Thus, for our scenario we transfer the burden of on-line segmentation to the involved

```
1  <htd:deadlines>
2    <htd:completionDeadline name="notifyManagement">
3      <htd:documentation xml:lang="en-US">notify the requester on deadline</htd:documentation>
4      <htd:for>htd:getInput("SLA")//wsla:Expression[wsla:SLAParameter='deadline']/wsla:value</htd:for>
5      <htd:escalation name="deadlineMissReview">
6        <htd:condition><![CDATA[htd:getInput("DocumentReview")/reviewComplete=true()]]></htd:condition>
7        <htd:notification name="deadlineMissSupervisor">
8          <htd:documentation xml:lang="en-US">inform the hierarchy of responsible roles</htd:documentation>
9          <htd:interface portType="tns:ReviewHandlerPT" operation="escalate" />
10         <htd:peopleAssignments>
11           <htd:recipients>
12             <htd:from logicalPeopleGroup="taskSupervisors">
13               <htd:argument name="supervisorID">
14                 htd:getActualOwner("AssignedSupervisor")
15               </htd:argument>
16             </htd:from>
17             <htd:from logicalPeopleGroup="taskCoordiantors">
18               <htd:argument name="coordinatorID">
19                 htd:getActualOwner("AssignedCoordinator")
20               </htd:argument>
21             </htd:from>
22           </htd:recipients>
23         </htd:peopleAssignments>
24         <htd:presentationElements/>
25       </htd:notification>
26     </htd:escalation>
27   </htd:completionDeadline>
28 </htd:deadlines>
```

Listing 4.4 Defined timeouts and escalation actions

crowd supervisors and hint a possible segmentation in this example by providing the number of chapters.

Deadlines. The WS-HT standard also provides sections to notify the associated parties in the process. The section related to a particular task is enclosed in the *deadlines* tag displayed in Listing 4.4. Related to our document review example, the defined notification chain is triggered by a completion deadline. The deadline itself has been agreed in the WSLA. An escalation is triggered if the condition stated is violated. In the example `DocumentReview`'s value `reviewComplete` is set to true if the complete review has been submitted to the aforementioned `result_url`. If the condition is broken then the assigned supervisors and coordinators are informed about the SLA violation.

4.5 Social Aggregator

Today, social networks are a mass phenomenon found in private (e.g., Facebook) and professional environments (e.g., LinkedIn). It is reasonable to assume that the trend of social networks will continue to penetrate more and more aspects of our lives. In a business context a social network is either formed explicitly, by manually adding contacts, or implicitly, based on observed interaction and collaboration patterns. We argue that it is highly beneficial to consider social aspects in crowdsourcing since a clique in a social network is more likely to efficiently work on collaborative tasks than a group of random workers. On a global scale, members of the latter are likely to have never worked together before, to have different cultural background, to speak a different language, and to live in different timezones, which altogether makes it extremely hard to create high-quality task processing results. The structure of companies is typically organized hierarchically. We borrow that concept to some extent and distinguish between three roles in our crowdsourcing system:

- *Workers* perform the actual processing of tasks and are assigned to one or multiple supervisors.
- *Supervisors* represent the head of a group of workers. They are responsible for breaking a task down into subtasks, to distribute those subtasks to suitable workers in her/his team, and to finally check the result. Each supervisor in turn is assigned to one or multiple coordinators.
- *Coordinators* are the interface between the customers who submit tasks to the social network and the supervisors who are responsible for tasks processing.

Figure 4.3 shows the main steps that are performed to augment the social network in order to make it suitable for crowdsourcing. The origin is a social network (c.f., Fig. 4.3a) that was formed either explicitly or implicitly; the nodes denote users of the crowdsourcing system and the edges social relationships between those users. Every user has a profile describing her/his skills. In the next step, illustrated in Fig. 4.3b, role hierarchies are formed.

We use *betweenness centrality*, a measure from graph theory indicating the importance of a node, to determine a member's role. We propose betweenness central-

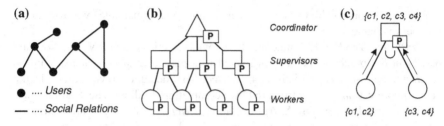

Fig. 4.3 Social roles and aggregation. **a** Social network. **b** Role hierarchies. **c** Aggregators

ity because it is often used in social and communication networks to estimate the potential monitoring and control capabilities a node may have on data flowing through the network [14]. In particular, we assume that nodes obtaining the role of a supervisor will have a high betweenness centrality value because these nodes have great influence on task flows. Let us define the graph $G(V, E)$ consisting of the set of vertices V and the set of edges E. Shortest-path betweenness centrality, as used in this work, defines the importance of a node s based on how many pairs of vertices go through $s \in V$ in order to connect through shortest paths in G (e.g., see [43]). Betweenness centrality $B(s)$ is formally defined as:

$$B(s) \leftarrow \sum_{u \neq s \neq t} \frac{g_{ut}(s)}{g_{ut}} \qquad (4.1)$$

where g_{ut} is the number of shortest paths linking nodes u and t; and $g_{ut}(s)$ is subset of those paths that contain node s. When the betweenness centrality of a user exceeds a certain threshold τ_S s/he has the prerequisites for becoming a supervisor, if it is greater than an even higher threshold value τ_C s/he could adopt the role of coordinator. This functionality is outlined in Algorithm 4.

When the importance values are calculated and a user fulfills the basic requirements for becoming a supervisor or coordinator two different approaches are supported how to actually decide on the roles:

- *Invitation*: The platform invites users exceeding a certain betweenness centrality threshold to adopt a hierarchically higher role. The user may accept or decline the offer.
- *Nomination*: The platform only nominates candidates for higher roles based on their importance in the social network. Users connected to the nominee can vote whether they support the candidate. Only with a certain minimum number of supporters the user is awarded the higher role. This prevents to assign high roles to users who have a high number of relationships and therefore a high importance indicator but whose relationships are mostly superficial and weak.

The final step to make the social network 'crowdsourcing-ready' is to create the higher-role profiles as an aggregation of all affiliated user profiles, as seen in Fig. 4.3c. This provides the basis for simple and rapid matching of tasks to competent groups in social crowdsourcing, as described in the following section.

Algorithm 4 Detecting roles of users

1: **input:** The social collaboration graph $G(V, E)$.
2: **output:** The set V containing workers $u \in V$ with different roles in the social network.
3: τ_S // threshold for supervisors
4: τ_C // threshold for coordinators
5: // simplified betweenness centrality evaluation of nodes in network G (for details see [9])
6: **for each** $v \in V$ **do**
7:   **for each** $n \in N(v)$ /* neighbors of v */ **do**
8:     **if** *shortest path through n* **then**
9:       // save distance
10:     **end if**
11:   **end for**
12: **end for**
13: **for each** *node along shortest path to v* **do**
14:   // from the most distant s to v
15:   **if** $s \neq v$ **then**
16:     increase $B(v)$
17:   **end if**
18: **end for**
19: **for each** $u \in V$ **do**
20:   **if** $B(u) \geq \tau_S$ **then**
21:     *isSupervisor*$(u, true)$
22:   **else if** $B(u) \geq \tau_C$ **then**
23:     *isCoordinator*$(u, true)$
24:   **end if**
25: **end for**

4.6 Task Segmentation and Matching

In this section we detail our approach for task processing in the crowd. First, we explain in detail how human tasks (as defined in the context of B4P) are passed from coordinators to supervisors and finally assigned to workers. As the next step, we introduce a ranking approach to select the best suited coordinator.

4.6.1 Hierarchical Crowd Activities

Crowd activities can be structured as hierarchies (see Fig. 4.4) using *parent* and *child* relations. Child activities specify the details with respect to the (sub-)steps in collaborations, for example, sub-activities in the scope of a parent activity. This allows for the refinement of collaboration structures as the demand for a new set of activities (e.g., performed by different people and services) increases. The need for the dynamic refinement of collaboration structures is especially required when people have limited experience (the history of performed activities) with respect to a given objective or goal [39]. Furthermore, some people tend to underestimate the scale and complexity of an activity; thus the hierarchical model enables the segmentation of activities into sub-activities, which can be, for example, delegated to other people.

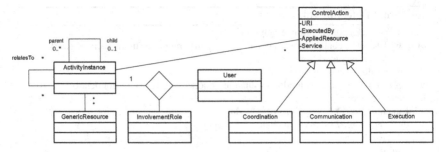

Fig. 4.4   Excerpt of hierarchical activity model

Activities have a *relatedTo* property which provides a mechanism to link to any other activity. Typically, multiple members work on the same activity with different roles. The *InvolvementRole* identifies the *coordinator*, *supervisor*, and responsible *worker* of an activity. Involved workers apply a set of *GenericResources* to perform their work. Objects such as documents are represented as a shared *Artifact*. A *ControlAction* captures activity-change events, interactions between members, and work carried out. Actions can trigger events describing the progress of activities.

4.6.2 Social Interactions

Figure 4.5 and related sub-figures show how task processing is performed in the crowd. We argue that interactions in crowdsourcing environments are governed by *user preferences*, *social trust*, and *reputation*.

Crowdsourcing follows the open world assumption which permits users to join and leave a particular community (platform) at any time. Instead of following a set of company rules or policies, crowd workers can be regarded as 'self-employed' individuals. However, it is in the interest of the worker to earn higher rewards and to work on tasks matching her/his expertise and interest. Also, we believe that complex tasks are typically rewarded higher as compared to simple tasks. Crowd members may decide to work with other members on a joint task based on previous experience or recommendations received from friends. Indeed, these interactions are not known in advance. Therefore, it is not possible to specify different task processing patterns that are performed in the crowd at the B4P or WS-HT level. The segmentation of human tasks is illustrated by Fig. 4.5.

As a first step in Fig. 4.5a we assume that the selected coordinator W1 forwards (e.g., through delegation) the human task to W5 who is a supervisor. The selection of the supervisor may entirely depend on W1's preferences to forward tasks to W5. Moreover, the previously discussed preferences as defined along with the groups may prevent W1 to forward a task to any of the supervisors s/he is connected to in the social network.

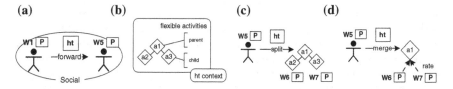

Fig. 4.5 Segmentation of tasks in social network. **a** Forward. **b** Crowd activities. **c** Assign. **d** Merge and rate

The supervisor receives a given task and performs some segmentation. In this context, we introduce *crowd activities*, which are collaborative activities performed by crowd members in a flexible manner. The notion of flexible (crowd) activities is independent of the previously discussed process activities that are designed in the context of processes such as BPEL. Here, we show how to combine *flexible* interactions and *top-down* process activities and tasks in order to support adaptive compositions of human- and software-based services. In our previous work [39] we have designed an activity model supporting collaborative working environments. The model and a set of collaboration tools have been implemented on top of Web services technology. Also, the inclusion of human capabilities in service-oriented collaboration environments is supported through the HPS concept [42]. These activities include, for example, 'write document', 'review document', 'proof-read paragraph'. Furthermore, these activities can be created and modified on-demand by people, e.g., the supervisor, based on their preferences and expertise in performing a specific type of task. Figure 4.5b shows an example of such an activity structure that can be created in a specific human task (ht) context.

Assume the 'review document' task (cf. Sect. 4.4) that needs to be outsourced to the crowd. The supervisor decides to split and to process the task by creating a hierarchical crowd-activity structure. A *parent* activity is initiated with the task's context data (presentation, elements, time constraints, etc.). Depending on the task's properties (e.g., duration, effort) sub-activity a2 and a3 are associated as *child* activities to a1. The segmentation step may be assisted by an *activity service* (a software service to manage crowd-activities) that recommends how many sub-activities the parent activity should be segmented to. Though, it is the responsibility of the supervisor W5 to allocate sub-activities to workers.

In social environments selection preferences and resulting interactions typically depend on the *trust* between actors. How much W5 trusts its neighboring peers (e.g., workers) is strongly influenced by previous interaction behavior. For example, W5 may trust a worker W6 more than W7 in performing a given activity depending on W5's collaboration experience. Positive experience results in higher trust between collaboration partners. We have established a set of metrics to measure collaboration experience (see [38, 40, 45]) including the *activity success* and *responsiveness* when processing an activity. A detailed description of these metrics and a trust model is, however, out of scope of this work. The assignment procedure is shown in Fig. 4.5c where W5 assigns a2 to W6 and a3 to W7. Each of these sub-activities can be controlled (e.g., inspecting the status and progress of an activity) by the supervisor. Once the workers W6 and W7 (see Fig. 4.5d) deliver the results, the supervisor takes the output of a2 and a3 and merges them. For example, the results can be combined by simply merging separate document sections to one document that was reviewed by W6 and W7. However, since the supervisor W5 is responsible for the final quality, W5 checks the result before the output of a1 is returned to the coordinator and/or B4P requester. How the result is being passed from the supervisor to the B4P requester may in fact depend on the 'social protocol' or preferences of actors. A coordinator may prefer to act as the main interface towards the crowd and thus may want to return the result.

The final step is the rating of the supervisor. Rating is performed to give feedback how well the supervisor distributes activities in the crowd. Crowd workers will be satisfied if the supervisor distributes activities that fit their expertise. Also, a worker's queue should neither be empty nor overloaded. This means that the supervisor should not accept too many tasks to avoid overload conditions. Careless assignments (e.g., activities that have low or no overlap with a worker's interest and skills) and false assumptions with regards to activity effort would cause bad ratings.

Next, we will introduce a ranking algorithm to rank coordinators based on capabilities and quality constraints specified by the B4P requester.

4.6.3 Ranking Coordinators

Here we introduce our novel ranking approach that bases its input on *skill information* as well as *social network metrics*. The approach consists of three essential steps that are briefly introduced in Algorithm 5.

Algorithm 5 Ranking approach outline

---

input: The social graph $G(V, E)$ and detected roles.
output: Ranked list of coordinators.

1. Calculate *importance scores* in the hierarchical social network (detailed by Algorithm 6).
2. Calculate the rank of supervisors *SR* based on

   - their skills and
   - their social standing (reputation) within the social network.

   Also, calculate the rank of *each worker* based on

   - their skills and
   - task load conditions.

   Append the workers' rank connected to a particular supervisor to *SR*.
3. Each coordinator gets the ranking scores of the top-ranked supervisor it is connected to. Sort coordinators according to their ranking score.

Let us start with the definition of a procedure to rank the importance of individuals in social networks. In this work we utilize the concept of hubs and authorities in Web-based environments. This concept has been introduced by Kleinberg [21] to rank Web pages in search queries using the hubs and authorities algorithm. The notion of authorities in social or collaborative networks can be interpreted as a measure to estimate the relative standing or importance of individuals in social networks. Compared to methods such as PageRank [31], the main advantage of the hubs and authorities model is that both hub and authority scores are calculated for each node in the network.

Applying this idea in our scenario, assume an undirected social network and roles of users that were detected using the previously introduced approach (cf. Algorithm 4). Also assume the hierarchical network that can be created considering the roles and the social graph. Coordinators are responsible for forwarding tasks to supervisors, thereby acting as *hubs* in the network. According to the theory developed in [21], the hub $H(u)$ and authority $A(u)$ scores of nodes $u \in V$ in the network is calculated as:

$$H(u) \leftarrow \sum_{(u,v)\in E} A(v) \qquad A(u) \leftarrow \sum_{(v,u)\in E} H(v) \tag{4.2}$$

However, we assume multiple roles in the social network such as coordinators, supervisors, and workers. Workers develop expertise in different areas depending on their interest and task processing behavior. Given the model in (4.2), workers would have higher authority scores if they receive requests to perform activities from supervisors that are connected to many 'good' authorities (i.e., workers). Good workers are characterized by their reliable task processing behavior which is monitored at runtime. Note, task processing behavior is observed through interaction monitoring techniques. Interaction metrics are established to obtain weighted social relations between actors. These relations are used to automatically calculate social network metrics such as hub- and authority-based importance scores.

Thus, a supervisor trusts a particular worker if the worker processes tasks in timely and satisfactory manner. Supervisors are rated by workers based on the suitability of assigned work. For example, supervisors who carelessly delegate tasks to workers without knowing their interests or who ignore the workers' load conditions (e.g., oversupply of task assignments in too short time frames) would receive bad ratings. Ratings thereby influence the weight of a relationship between worker and supervisor. Notice, both trust and rating relations are established upon interaction monitoring and mining. Thus, in addition to the undirected social links, directed relations are introduced based on collaborations between actors in the crowdsourcing environment. The recursive definition of hub and authorities is typically computed using an iterative algorithm. In the Algorithm 6, we introduce an extended hubs and authorities algorithm suitable for calculating hub- and authority scores in hierarchical social networks.

The goal of the algorithm is to calculate:

- hub scores for coordinators as they forward tasks to supervisors through delegation actions,
- hub scores for supervisors as they perform pre-processing of tasks and create flexible crowd-activities that are distributed and assigned to workers,
- authority scores for supervisors as they receive task requests from coordinators, and
- authority scores for workers as they perform the actual work.

Algorithm 6 Hubs and authorities algorithm in hierarchical social networks

---

1: // Initialize hub and authority scores
2: **for** $u \in V$ **do**
3:    $\mathcal{H}(u)^{(0)} \leftarrow p_u^H, \mathcal{A}(u)^{(0)} \leftarrow p_u^A$
4: **end for**
5: $t = 1$
6: **while** not converged **do**
7:    **for each** $u \in V$ **do**
8:       // update ranking scores
9:       **for** $v \in N(u)$ **do**
10:          **if** $isCoordinator(u)$ **then**
11:             $\mathcal{H}(u)^{(t)} \leftarrow \mathcal{H}(u)^{(t-1)} + \mathcal{A}(v)^{(t-1)}$
12:          **else if** $isSupervisor(u) \wedge \neg isCoordinator(v)$ **then**
13:             $\mathcal{H}(u)^{(t)} \leftarrow \mathcal{H}(u)^{(t-1)} + w_{vu}\mathcal{A}(v)^{(t-1)}$
14:          **else if** $isSupervisor(u) \wedge isCoordinator(v)$ **then**
15:             $\mathcal{A}(u)^{(t)} \leftarrow \mathcal{A}(u)^{(t-1)} + \mathcal{H}(v)^{(t-1)}$
16:          **else if** $isSupervisor(v)$ **then**
17:             $\mathcal{A}(u)^{(t)} \leftarrow \mathcal{A}(u)^{(t-1)} + w_{vu}\mathcal{H}(v)^{(t-1)}$
18:          **end if**
19:       **end for**
20:       // Normalize rankings and test for convergence
21:       $t = t + 1$
22:    **end for**
23: **end while**

The first step in Algorithm 6 (see lines 2–4) is to initialize two vectors $\mathcal{H}$ and $\mathcal{A}$ that hold hub and authority scores, which are updated after each iteration t. Without any prior (node bias), the initialization vectors p^H and p^A hold for each node the same initial value. The main loop (lines 6–23) is executed until the ranking scores converge (i.e., the ranking order of nodes is no longer changed between the step $t-1$ and t). For each node $u \in V$, we update hub- and authority scores according to the aforementioned update procedure. The set of u's neighbors is obtained by $N(u)$. Assume that u is a coordinator (line 10), then only u's hub score $\mathcal{H}(u)$ is updated. The next case holds (line 12) if u is a supervisor and the neighbor v is a worker. In this case, the supervisor's hub score needs to be updated.

As mentioned before, the hub score of supervisors is influenced by ratings they receive from workers. Thus, the authority score of v is added with weight w_{vu} to $\mathcal{H}(u)$

(see line 13). In case v holds the role of a coordinator (line 14), u's authority score is updated. Notice, weights are only calculated between supervisors and workers as we assume stronger collaborations between these two actors whereas coordinators mainly act as 'entry points' to the crowdsourcing platform.

The final procedure (line 16–18) is performed to update the worker's authority score. Here, the score $\mathscr{H}(v)$ is appended with weight w_{vu} that is calculated based on mining metrics (e.g., how much the supervisor v trusts the worker u). After these steps, a check in line 20 verifies if convergence has been reached. For larger social networks a fixed number of steps can be used to reduce the time needed for computing importance scores. After convergence the final scores are copied into H and A.

Next, we introduce the ranking procedure to process crowd-activities that can be segmented into flexible activity structures. Certain activities can be decomposed hierarchically into sub-activities depending on their required processing effort or complexity. Algorithm 7 shows the procedure to rank coordinators. As input we assume the social network graph $G(V, E)$, an activity $a \in A$ which could be part of a complex activity structure, and the set V^C of coordinators. The goal of the algorithm is to assign a ranking score to each user $u \in V^C$.

Algorithm 7 Rank coordinators based on social graph G

1: **input:** $G(V, E)$ representing the social network graph, splittable activity $a \in A$ and the set of coordinators V^C that have already been filtered based on *hard constraints*.
2: **output:** Ranked list of coordinators (CR).
3: **for** $u \in V^C$ **do**
4:   // Get supervisors connected to u
5:   **for each** $v \in N(u)$ **do**
6:     **if** $isSupervisor(v)$ **then**
7:       $V_u^S \leftarrow V_u^S \cup v$
8:     **end if**
9:   **end for**
10:   **for each** $v \in V_u^S$ **do**
11:     // Get workers connected to v
12:     **for** $n \in N(v)$ **do**
13:       **if** $\neg isCoordinator(n)$ **then**
14:         $V_v^W \leftarrow V_v^W \cup n$
15:       **end if**
16:     **end for**
17:     $SR(v, a) \leftarrow \alpha \cdot skill(v, a) + (1 - \alpha) \cdot (0.5 \cdot A(v) + 0.5 \cdot H(v))$
18:     **for each** $n \in V_v^W$ **do**
19:       $score_n \leftarrow \beta \cdot skill(n, a) + (1 - \beta) \cdot (1 - getRate(n, a, DueAt(a))$
20:       $score_{V_v^W} \leftarrow score_{V_v^W} + \frac{1}{|V_v^W|} score_n$
21:     **end for**
22:     $SR(v, a) \leftarrow SR(v, a) + score_{V_v^W}$
23:   **end for**
24:   $s \leftarrow getTopRanked(SR(a))$
25:   $CR(u, a) \leftarrow getScore(s, a)$
26: **end for**
27: // Order by *soft constraints*: sort CR in descending order
28: **return** CR

Table 4.1 Description of calculations and equations

Ranking input	Description
Skill	The skill of supervisors and workers. The function $skill(v, a)$ returns v's skill level with respect to an activity a. For example, if a demands for language skills in English with the level 'expert' and v's experience in English is at the level 'expert', v's skills match perfectly a's skill requirements. Also, skill profiles are automatically maintained by updating the users' experience. In our previous work we have designed and implemented algorithms for profile matching [41] and skill updates [37]. Details regarding skill matching and update are not presented in this work.
Importance	The relative standing of a user within the social network. As explained before, the importance of a node is based on the concept of hubs and authorities in social networks. The supervisor's importance is determined by both hub and authority scores due to the hierarchical nature of the previously explained social (collaborative) network. Hence, v's importance score is the weighted sum of the authority score $A(v)$ and the hub score $H(v)$. Different weights could be used to assign preferences to either score.

First, a set V_u^S is initialized that contains supervisors connected to u (see lines 5–9). The next loop (lines 10–23) shows how to calculate rankings scores for supervisors. Note that the coordinator u acts as a 'proxy' for supervisors; thus u's score is based on the score of the highest ranked supervisor (lines 24–25). The basic idea to calculate rankings for supervisors has been shortly explained in Algorithm 5. Our approach (see Algorithm 7) is to rank supervisors based on the actual (observed) skills (line 17), the importance of supervisors in the social networks (line 17), and the actual skills of workers a supervisor is connected to (lines 18–21). The set of workers V_v^W connected to v is first initialized (lines 12–16). In the following (lines 17–22) the computation of v's ranking score is shown. The initial supervisor ranking score SR is calculated as shown in line 17. The initial score is the aggregation (weighted by the parameter α) of the supervisor's skills and (social) importance. These input parameters are detailed in the following Table 4.1.

The next steps (lines 18–21) in Algorithm 7 is to calculate ranking scores for each worker connected to the supervisor v. This is done in a similar manner as for the supervisor. However, instead of considering the importance of a worker within the social network, we take other workload related factors into account. The function *getRate* calculates the workers' rate based on the earliest possible start time (influence by the workers' task queue size) and activity effort. This means that even if a supervisor has high skills and high importance, it still needs to be connected to a set of workers who have free resources in terms of free time to process a crowd activity. Otherwise, the supervisor would need to handle all activities him/herself. The score of each worker is appended with equal weight $\frac{1}{|V_v^W|}$. The final score SR is the sum of the supervisor's initial ranking score plus the workers' ranking scores.

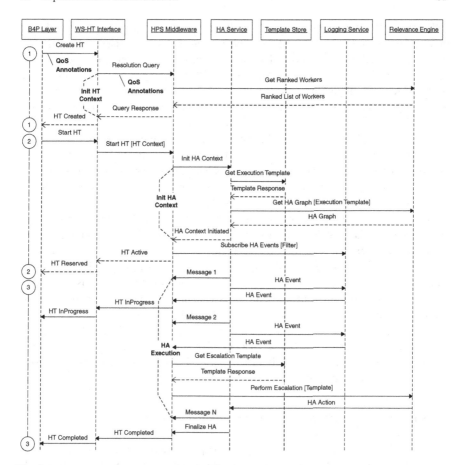

Fig. 4.6 Sequence diagram of adaptive B4P process execution

4.7 Implementation and Evaluation

Our evaluation and results are based on a proof of concept implementation of various introduced concepts and simulations of interactions in social-crowd environments. The following Sect. 4.7.1 describes the SOA-based crowdsourcing environment including the lifecycle of a human task and the principle interactions between services, Sect. 4.7.2 explains how the basic social network structure has been generated and Sect. 4.7.3 presents our findings.

4.7.1 SOA-Based Crowdsourcing Environment

In this section we provide an overview of the main services and the most important interactions between services (see Fig. 4.6). The implementation of our NFP-aware

B4P execution environment is mainly built on-top of a service-oriented collaboration environment. The collaboration services, however, can be used independently of any top-down process model. The main extensions of the environment consist of the WS-HT Interface (a plugin of the HPS Middleware) to provide a bridge between B4P and the crowdsourcing environment. The protocol between the B4P Layer and the WS-HT Interface is in conformance with the WS-HT [4] standard.

The collaboration environment consists of a SOA-runtime for mixed service-oriented systems (see HPS Middleware). Unlike traditional SOA-based systems, also human-based services (i.e., HPSs) are made available for discovery and invocation [42]. Coordination and collaboration among people and services (HPS and SBS) is achieved by using an activity service (HA Service). The Template Store contains activity skeletons (e.g., activity structure) that can be instantiated at runtime. Such templates include, for example, the definition of parent child activities to perform a document review. The Logging Service monitors all interactions and saves XML-messages and additional metadata in a database for later analysis. The Relevance Engine implements ranking and mining algorithms.

The lifecycle of human task execution is structured into three essential phases. First, a resolution query is performed to find suitable candidate workers who can process a human task. Second, a crowd-activity structure is initialized that allows crowd-members to process activities in a flexible manner. Third, workers collaborate to jointly work on activities (collaboration phase). Figure 4.6 details the interactions between the various services.

4.7.1.1 Human Task Creation and Resolution of Workers

A request to create a human task that is to be performed by the crowd is initiated by the B4P Layer. This layer is typically implemented as an extension of a BPEL orchestration engine. The specification of a human task contains additional elements to ensure the quality of a task's result (cf. **QoS Annotations**). These annotations have been introduced in the context of Listing 4.3 and define the required set of human capabilities, which are matched against capability profiles, and the required quality. NFP elements such as human capabilities are used in the matching procedure (see arrow *Resolution Query*).

```
1   <mex:Metadata>
2     <mex:MetadataSection Dialect="http://schemas.xmlsoap.org/wsdl/">
3       <wsdl:definitions>
4         <!-- Omitted -->
5       </wsdl:definitions>
6     </mex:MetadataSection>
7     <mex:MetadataSection Dialect="http://xmlns.com/foaf/0.1/">
8       <rdf:RDF xmlns:foaf = "http://..." xmlns:capability = "http://.../capability.owl#">
9         <foaf:Person rdf:about="http://www.tuwien.../staff/">
10          <foaf:name>H. Psaier</foaf:name>
11          <foaf:interest rdf:resource="http://.../hpsaier/interests.rdf"/>
12          <!-- Omitted -->
13          <capability:op>
```

```
14      <capability:port id="TSportType">
15        <capability:op id="translateDoc">
16          <capability:opwsdlxpath>
17            wsdl:operation/[@name="TSportType"]
18          </capability:opwsdlxpath>
19          <capability:opmetricgrounding
20            rdf:resource="http://.../grounding-translateDoc.xml"/>
21            <capability:opmetric>
22              <capability:opmetricid>cost</capability:opmetricid>
23              <capability:opmetricvalue>100.0</capability:opmetricvalue>
24            </capability:opmetric>
25            <capability:opmetric>
26              <capability:opmetricid>reliability</capability:opmetricid>
27              <capability:opmetricvalue>0.8</capability:opmetricvalue>
28            </capability:opmetric>
29              ...
30        </capability:op>
31      </capability:port>
32    </foaf:Person>
33    </rdf:RDF>
34  </mex:MetadataSection>
35 </mex:Metadata>
```

Listing 4.5 HPS metadata exchange description

Listing 4.5 shows the simplified structure of the resolved HPS information. NFP elements are embedded in the HPS's WSDL interface. In addition, an extended FOAF description is inserted into a *WS-Metadata-Exchange*[10] (MEX) document (see also [42]). The HPS framework uses SPARQL to define search queries[11] on FOAF structures. The sample response message to a MEX GET request in Listing 4.5 comprises the following elements. The main response body contains the currently offered operations in a WSDL (omitted for brevity) and the related NFPs in the second `MetadataSection` in FOAF format. The elements with the capability prefix provide the current NFP values for a related operation defined in the WSDL section. In our current implementation, such NFPs are costs and primarily quality metrics, such as the HPSs reliability and responsiveness. The XPath statement identifies an operation uniquely. The following metric grounding resource `opmetricgrounding` links a document with metric definitions (meaning, measurement, unit, range of values, etc.) to the listed metric ids. The `HPS Middleware` interacts with the `Relevance Engine` to obtain a ranked list of workers. For simplicity, we do not discuss the different social roles such as coordinators or supervisors in this context. Notice, the result of a resolution query is a list of coordinators if the task can be segmented in multiple crowd-activities. The successful result of this interaction is denoted by the arrow *HT Created*.

4.7.1.2 Reserve Human Task and Initialize Activity Structure

The activity structure is being initialized by *Start HT*. The `WS-HT Interface` passes the HT Context to the `HPS Middleware`, which in turn signals *Init HA*

---

[10] http://www.w3.org/Submission/WS-MetadataExchange/

[11] http://www.w3.org/TR/rdf-sparql-query/

Context to the HA Service. Depending on the selected HT Context, different activity execution templates can be selected (*Get Execution Template*). An execution template may define how activities are processed. For example, if the result that is provided in the context of a specific human task has always low quality, an additional quality assurance step can be inserted dynamically in the execution template. The next step is to assign people to activities that are part of the execution template (see *Get HA Graph*). Ranking of people is performed by the Relevance Engine[12] (cf. to discussions related to matching and ranking in the previous section). The Logging Service logs all service interactions (i.e., SOAP calls) and also events triggered by the activity service. Activity events are fired based on activity changes (start, suspend, or finalize activity) and actions taken by human actors. Such actions include delegations of activities or the assignment of new activities. The Logging Service implements a publish/subscribe mechanisms that allows subscribers to get notified about specific events. The HPS Middleware subscribes to activity change events to monitor the status of activities (see arrow *Subscribe HA Events*). The result of these steps is *HT Reserved*.

4.7.1.3 Task Execution and Escalation Handling

In service-oriented systems, people interact and collaborate by using tools and services to perform their work. Each service call (performed in the context of an activity) is processed by the HPS Middleware. The middleware implements a SOAP dispatcher that performs message inspection and routing. The HA Service notifies the Logging Service about activity changes (see *HA Event*). Here the activity status is changed to 'activity in-progress'. The event is also sent to the middleware which signals *HT InProgress*. A series of messages 1...N is then exchanged between the HA Service and the HPS Middleware until an activity is finalized. Escalations are defined in the context of a human task (cf. Listing 4.4). As mentioned before, the HPS Middleware acts as a bridge between the B4P-based process, the activity-based collaboration services and tools that are used by crowd workers. Thus, the middleware monitors the status of activities and checks whether deviations in the progress of activities may cause deadline violations. The Relevance Engine receives a *Perform Escalation* call to trigger a *HA Action* if a deadline is going to be violated. As shown previously in Listing 4.4, a notification may be the result of such an escalation action. The Relevance Engine performs the escalation by sending the *HA Action* to the activity service. Notice, escalations are not directly performed by the HPS Middleware. The Relevance Engine deals with escalations to support dynamic aspects (e.g., adaptive notification chains) and also future extensions of our approach such as complex event processing features. *HT Completed* is triggered once *Finalize HA* is received from the activity service.

---

[12] The Relevance Engine has by default access to all logs and events collected in the environment.

4.7.2 Social Network Generation

In our experiments, we generate synthetic social graphs to test the applicability and effectiveness of our proposed ranking model. At the time when performing this research, a sufficiently large crowd user-base was not available to perform tests with real users. We use two different methods to generate social graphs: *random graphs* [28, 30] are generated and graphs based on the *preferential attachment model* [6, 35]. The more general case are random graphs wherein each pair of nodes has an identical, independent probability of being joined by an edge. Preferential attachment results in more specific graphs wherein nodes preferentially connect to existing nodes with high degree (the 'rich get richer'). By using these two methods, we are able to evaluate the effectiveness of our ranking approach by considering different social network structures. Figure 4.7 shows a basic social network structure that has been generated according to the statistical properties as found in freely emerging networks. Each figure visualizes a graph with 200 workers.

1. *Random graphs* are based on the assumption that any random actor will establish a connection to some other random actor with probability p. The resulting graph structure is visualized by Fig. 4.7a. In our experiments, we use a probability of 0.3 that an actor u will establish a connection with a random actor v.
2. *Preferential attachment graphs* are based on the assumption that networks emerge according to the rule of preferential attachment [35]. This process produces a scale-free graph with node degrees following a power-law distribution. The resulting social graph represents very well the structure of autonomously forming collaborations in cooperation networks [29].

By using a probability of 0.3 to generate random graphs, both graphs, random and preferential, have approximately the same amount of edges; thereby making the both types of graphs comparable with regards to number of workers and number of edges. Roles in the social network were detected according to Algorithm 4. Coordinators are visualized as triangular shapes, supervisors are depicted by rectangles, and regular workers are shown as circular nodes. One can see that the random graph in Fig. 4.7a exhibits only sparsely connected nodes when compared to Fig. 4.7b. Using these two graphs, we are able to compare the results of our ranking approach under different conditions. This is an important issue because sparse networks are a natural phenomenon in newly established social networks.

In each network, workers have certain skills associated with it. In our experiments we only use a single skill whose skill level is distributed according to a normal distribution $\mathcal{N}(\mu, \sigma^2)$ with a mean value $\mu = 0.6$ and a standard deviation $\sigma^2 = 0.25$. The parameters of this model (mean value and standard deviation) yield the following skill level properties of the resulting worker population: most workers have good skills in performing their tasks with an average skill level of 0.6, some workers are highly skilled with a maximum skill level of 1.0 (top expert) and on the contrary some workers have a very low skill level (in our experiments the minimum skill level was 0.02). If a higher or lower average value would be chosen, the expected quality

(a)   **(b)**

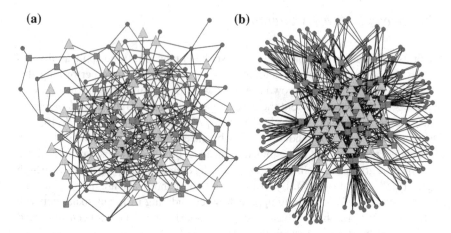

Fig. 4.7 Generated social graphs: **a** sparsely connected random graph. **b** preferential attachment graph

Table 4.2 Configurations for different experiments

Configuration	1	2	3	4	5	6
Number of workers	100	100	100	100	200	200
Activities per round	5	5	10	10	5	5
Advanced processing	No	Yes	No	Yes	No	Yes

of returned tasks can also be expected to be higher or lower respectively. If a higher standard deviation is chosen, the likelihood of having more highly skilled workers as well as workers with very low skills increases. By choosing a lower standard deviation, it is more likely that the workers will have the average skill level of 0.6 and it is less likely that workers have high or low skills.

4.7.3 Discussion

We performed several experiments and compared the quality of task results considering task processing with and without social network structures. The default option of our simulation is to process activity in the context of a human task without advanced processing. This configuration provides the baseline results for comparison with the advanced processing option. The configurations of our experiments are detailed in Table 4.2. The entry *advanced processing* indicates whether certain activities were split and processed collaboratively in social networks.

Table 4.2 shows three pairs of experiments (1, 2), (3, 4), and (5, 6). Each pair compares the default processing behavior with the advanced processing option. Advanced processing means that actors' behavior is guided by their social role. Coordinators forward task requests to supervisors which split tasks into multiple (crowd-) activities

Table 4.3 Numerical values of experiment results using random graph

Configuration	1	2	3	4	5	6
Created activities	1000	2237	2000	4106	1000	2208
Finished activities	940	2234	1147	3950	989	2108
Average quality	0.720	0.736	0.488	0.607	0.847	0.907
Overdue activities (%)	23	2	13	1	7	1

(a)

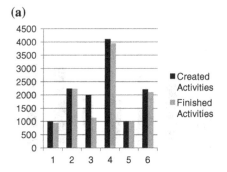

(b)

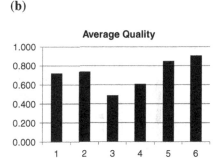

(c)

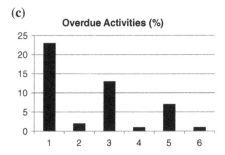

Fig. 4.8 Experiment results using random graph. **a** Activity creation. **b** Activity quality. **c** Activities overdue

that are assigned to workers. In our simulation, tasks are issued by the B4P requester in fixed rounds. In each round, 5 tasks are issued in configuration 1 and 2 and also in 5 and 6. The configurations 3 and 4 are based on 10 tasks per round to analyze processing behavior (e.g., quality) under different load conditions.

4.7.3.1 General Case: Random Graphs

The first set of experiments were performed using random graphs as depicted in Fig. 4.7a. However, we vary the number of workers according to the previously described configurations.

Table 4.3 shows the numerical results, which are visualized in Fig. 4.8.

Table 4.4 Numerical values of experiment results using preferential attachment graph

Configuration	1	2	3	4	5	6
Created activities	1000	2237	2000	4406	1000	2208
Finished activities	944	2233	1258	4062	989	2208
Average quality	0.724	0.799	0.492	0.550	0.847	0.873
Overdue activities (%)	22	1	12	1	6	0

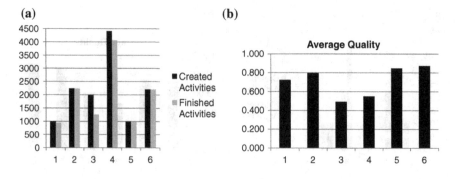

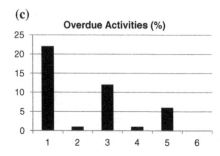

Fig. 4.9 Experiment results using preferential attachment graph. **a** Activity creation. **b** Activity quality. **c** Activities overdue

4.7.3.2 Specific Case: Preferential Attachment Graphs

The second set of experiments were performed using preferential attachment graphs as depicted in Fig. 4.7b. Again, we vary the number of workers according to the previously described configurations.

The Table 4.4 shows the numerical results, which are visualized in Fig. 4.9.

4.7.4 Overall Findings

Both sets of figures, Figs. 4.8 and 4.9 show the results of our experiments by comparing the different pairs of configurations. The horizontal axis of each figure shows the index of a configuration that corresponds to the simulation parameters as defined in Table 4.2. In general, both graphs (random and preferential attachment) exhibit similar results with only minor differences. This means that our proposed ranking approach is applicable to both, sparsely connected random graphs as well as more densely connected preferential attachment graphs. Thus, the following discussions apply to both sets of experiments using respective graph structure.

The first series of experiments shows the relation of the number of created activities versus the number of finished activities. Without advanced processing, an activity is simply created based on the properties of a human tasks and assigned to individual workers. On the other hand using advanced processing, if the duration of a task exceeds a certain duration threshold, an activity is created that is split into multiple sub-activities. The supervisors distributes sub-activities in the context of a parent activity, assembles the result, and passes it on to the coordinator.

Both Figs. 4.8a and 4.9a show that the number of activities is always higher in social-crowd environments (i.e., advanced processing) because activities are split and reassigned to workers. However, the number of finished activities in relation to the number of created activities is always higher when compared to the regular processing behavior. This means that advanced processing increases the number of created *and* successfully finished activities (i.e., the reliability in processing activities in crowdsourcing environments increases).

Figures 4.8b and 4.9b visualize the average quality obtain in different experiment configurations. The quality of a task result is based on the worker's skill (regular processing) or the supervisor's skill (advanced processing). Thus, in the latter case the quality is ensured by the supervisor. The average quality of tasks is always higher in the advanced processing case. This is the result of our ranking approach which ensures that coordinators are ranked higher if they are connected to skilled supervisors. Comparing the pairs of configurations, the quality in the configuration pair 3 and 4 is lower due to the larger number of activities to be processed. However, our advanced processing approach still outperforms the regular processing setting in terms of providing better quality results. Also, given a larger social network of 200 workers the task quality is higher.

Finally, Figs. 4.8c and 4.9c show the number of overdue activities which were not processed on time (deadline violations). The percentage ratio of overdue activities is much lower in the social-crowd environment because larger tasks (based on effort/-duration of a task) are split into smaller crowd-activities which are processed faster than larger chunks of work. It is easier to assign smaller tasks to crowd members instead of finding people to process larger tasks; thereby reducing the number of overdue activities. To conclude our discussions, we confirm that the proposed social-crowd environment has a number of advantages over traditional environments that are based on a population of workers which perform tasks separately. Our experiments show that task quality is increased while improving reliability and performance of the crowd.

4.8 Conclusion and Future Work

Crowdsourcing has emerged as an important paradigm in human problem solving techniques on the Web. In such environments, people offer their skills and capabilities in a service-oriented manner. However, one cannot rely on the constant availability of people. The dynamic discovery of skilled people becomes a key aspects. Here we proposed social-crowds that collaboratively process tasks. We designed extensions for BPEL4People to utilize crowds in process-centric enterprise environments. We explained in detail various extensions to cope with quality issues. Furthermore, we proposed a role detection algorithm to build up hierarchical social networks. The presented social-crowd environment brings a number of benefits including (i) increased task quality and (ii) an increased number of successfully finished activities as well as (iii) a reduced number of overdue activities. We believe that social-crowd environments have a great potential to make crowdsourcing more reliable while increasing quality of task results.

Task costs in crowdsourcing have not been detailed in this work (see our previous work in [34, 37]) but will be addressed in the context of B4P in future work. We plan to utilize AMT for experiments with real people and we will investigate the integration of various XML-based standards and interfaces including B4P, WS-HT, and AMT's API. Also, our future work will specifically deal with the question of stakeholder support in the context of BPM. In particular, the question we attempt to answer is which stakeholders need to be involved when designing novel crowdsourcing applications. For example, engineers may be interested in dynamic interaction and discovery policies whereas business analysts may want to design different incentive schemes for crowdsourcing services. These questions have not been addressed in our current research.

References

1. Adams, M., ter Hofstede, A.H.M., Edmond, D., van der Aalst, W.M.P.: Worklets: a service-oriented implementation of dynamic flexibility in workflows. In: OTM Conferences vol. 1, pp. 291–308, 2006
2. Agichtein, E., Castillo, C., Donato, D., Gionis, A., Mishne, G.: Finding high-quality content in social media. In: WSDM, pp. 183–194. ACM (2008)
3. Agrawal, A., et al.: Ws-bpel extension for people (bpel4people), version 1.0, 2007
4. Amend, M., et al.: Web services human task (ws-humantask), version 1.0, 2007
5. Balthazard, P.A., Potter, R.E., Warren, J.: Expertise, extraversion and group interaction styles as performance indicators in virtual teams: how do perceptions of it's performance get formed? Database 35(1), 41–64 (2004)
6. Barabasi, A.L., Albert, R.: Emergence of scaling in random networks. Science 286(5439), 509–512 (1999)
7. Benkler, Y.: Coase's penguin, or linux and the nature of the firm. CoRR, cs.CY/0109077 (2001)
8. Brabham, D.: Crowdsourcing as a model for problem solving: an introduction and cases. Convergence 14(1), 75 (2008)
9. Brandes, U.: A faster algorithm for betweenness centrality. J. Math. Sociol. 25, 163–177 (2001)

10. Breslin, J., Passant, A., Decker, S.: Social web applications in enterprise. Soc. Semant. Web **48**, 251–267 (2009)

11. Cozzi, A., Farrell, S., Lau, T., Smith, B.A., Drews, C., Lin, J., Stachel, B., Moran, T.P.: Activity management as a web service. IBM Syst. J. **45**(4), 695–712 (2006)

12. Cugola, G., Nitto, E.D., Fuggetta, A., Ghezzi, C.: A framework for formalizing inconsistencies and deviations in human-centered systems. ACM Trans. Softw. Eng. Methodol. **5**(3), 191–230 (1996)

13. Doan, A., Ramakrishnan, R., Halevy, A.Y.: Mass collaboration systems on the world wide web. Commun. ACM **54**(4), 86–96 (2011)

14. Dolev, S., Elovici, Y., Puzis, R.: Routing betweenness centrality. J. ACM **57**, 25:1–25:27 (2010)

15. Easley, D., Kleinberg, J.: Networks, Crowds, and Markets: Reasoning About a Highly Connected World. Cambridge University Press, Cambridge (2010)

16. Gentry, C., Ramzan, Z., Stubblebine, S.: Secure distributed human computation. In: EC '05, pp. 155–164. ACM (2005)

17. Herrmann, K., Rothermel, K., Kortuem, G., Dulay, N.: Adaptable pervasive flows—an emerging technology for pervasive adaptation. In: Workshop on Pervasive Adaptation (PerAda), Sept. 2008

18. Howe, J.: The rise of crowdsourcing. http://www.wired.com/wired/archive/14.06/crowds.html, June 2006

19. IBM.: An architectural blueprint for autonomic computing (whitepaper), 2005

20. Ipeirotis, P.G.: Analyzing the amazon mechanical turk marketplace. SSRN eLibrary **17**(2), 16–21 (2010)

21. Kleinberg, J.: Authoritative sources in a hyperlinked environment. J. ACM **46**(5), 604–632 (1999)

22. Kleinberg, J.: The convergence of social and technological networks. Commun. ACM **51**(11), 66–72 (2008)

23. Kumar, A., Aalst, W.M.P.V.D., Verbeek, E.: Dynamic work distribution in workflow management systems: how to balance quality and performance. J. Manag. Inf. Syst. **18**(3), 157–193 (2002)

24. Lampe, C., Resnick, P.: Slash(dot) and burn: distributed moderation in a large online conversation space. In: Proceedings of the SIGCHI Conference on Human Factors in Computing Systems, CHI '04, pp. 543–550. ACM, New York (2004)

25. Liu, L., Thanheiser, S., Schmeck, H.: A reference architecture for self-organizing service-oriented computing. In: ARCS, pp. 205–219, 2008

26. Maximilien, E.M., Singh, M.P.: Toward autonomic web services trust and selection. In: ICSOC '04, pp. 212–221. ACM (2004)

27. Mendling, J., Ploesser, K., Strembeck, M.: Specifying separation of duty constraints in bpel4people processes. In: BIS'08, pp. 273–284. Springer (2008)

28. Newman, M.E., Strogatz, S.H., Watts, D.J.: Random graphs with arbitrary degree distributions and their applications. Phys. Rev. E. Stat. Nonlin. Soft Matter Phys. 64(2 Pt 2), 026118 (2001)

29. Newman, M.E.J.: The structure of scientific collaboration networks. Proc. Natl. Acad. Sci. U. S. A **98**, 404–409 (2001)

30. Newman, M.E.J., Watts, D.J., Strogatz, S.H.: Random graph models of social networks. Proc. Natl. Acad. Sci. U. S. A. **99**(Suppl 1), 2566–2572 (2002)

31. Page, L., Brin, S., Motwani, R., Winograd, T.: The PageRank Citation Ranking: Bringing Order to the Web. Technical report, Stanford Digital Library Technologies Project (1998)

32. Panteli, N., Davison, R.: The role of subgroups in the communication patterns of global virtual teams. IEEE Trans. Prof. Commun. **48**(2), 191–200 (2005)

33. Petrie, C.: Plenty of room outside the firm. Internet Comput. **14**, 92–96 (2010)

34. Psaier, H., Skopik, F., Schall, D, Dustdar, S.: Resource and agreement management in dynamic crowdcomputing environments. In: EDOC, 2011

35. Reka, A., Barabási, A.-L.: Statistical mechanics of complex networks. Rev. Mod. Phys. **74**, 47–97 (2002)

36. Russell, N., Aalst, W.M.P.V.D.: Evaluation of the bpel4people and ws-humantask extensions to ws-bpel 2.0 using the workflow resource patterns. Technical report, BPM Center Brisbane/Eindhoven, 2007

37. Satzger, B., Psaier, H., Schall, D., Dustdar, S.: Stimulating skill evolution in market-based crowdsourcing. Springer, In: BPM, Lecture Notes in Computer Science 2011

38. Schall, D.: Human interactions in mixed systems—architecture, protocols, and algorithms. Ph.D. thesis, Vienna University of Technology (2009)

39. Schall, D.: A human-centric runtime framework for mixed service-oriented systems. Distrib. Parallel Databases **29**, 333–360 (2011). doi:10.1007/s10619-011-7081-z

40. Schall, D.: Expertise ranking using activity and contextual link measures. Data Knowl. Eng. **71**(1), 92–113 (2012). doi:10.1016/j.datak.2011.08.001

41. Schall, D., Skopik, F., Dustdar, S.: Expert discovery and interactions in mixed service-oriented systems. IEEE Trans. Serv. Comput. **71**(1), 233–245 (2012). doi:10.1109/TSC.2011.2

42. Schall, D., Truong, H.-L., Dustdar, S.: Unifying human and software services in web-scale collaborations. IEEE Internet Comput. **12**(3), 62–68 (2008). doi:10.1109/MIC.2008.66

43. Shi, X., Bonner, M., Adamic, L.A. Gilbert, A. C.: The very small world of the well-connected. In: HT '08, pp. 61–70. ACM (2008)

44. Siorpaes, K., Simperl, E.: Human intelligence in the process of semantic content creation. World Wide Web **13**, 33–59 (2010). doi:10.1007/s11280-009-0078-0

45. Skopik, F., Schall, D., Dustdar, S.: Modeling and mining of dynamic trust in complex service-oriented systems. Inf. Syst. **35**, 735–757 (2010)

46. Su, Q., Pavlov, D., Chow, J.-H., Baker, W.C.: Internet-scale collection of human-reviewed data. In: WWW '07, pp. 231–240. ACM (2007)

47. Thomas, J., Paci, F., Bertino, E., Eugster, P.: User tasks and access control over web services. In: ICWS '07, pp. 60–69. IEEE (2007)

48. von Ahn, L.: Games with a purpose. IEEE Comput. **39**(6), 92–94 (2006)

49. Vukovic, M.: Crowdsourcing for enterprises. In: Proceedings of the 2009 Congress on Services, pp. 686–692. IEEE Computer Society (2009)

50. Yang, J., Adamic, L., Ackerman, M.: Competing to share expertise: the taskcn knowledge sharing community. In: International Conference on Weblogs and Social Media, 2008

51. Zhang, J., Ackerman, M.S., Adamic, L.: Expertise networks in online communities: structure and algorithms. In: WWW, pp. 221–230. ACM (2007)

52. Zhao, X., Liu, C., Sadiq, W., Kowalkiewicz, M., Yongchareon, S.: Implementing process views in the web service environment. World Wide Web **14**(1), 27–52 (2011)

Chapter 5
Conclusion

The Web is evolving rapidly by allowing people to publish information and services. At the heart of this trend, interactions become increasingly complex and dynamic spanning both humans and software services. Thus, there has been a growing interest in the complex structure and dynamics of todays society. Our online-society is increasingly influenced by networks, incentives, and the behavior of social communities.

In this book, we analyzed the basic marketplace statistics of Amazon Mechanical Turk and derived a model for clustering tasks and requesters. Furthermore, we introduced a novel community discovery and ranking approach for task-based crowdsourcing markets. We have discussed a broker discovery and ranking model that lets other requesters discovery intermediaries who can crowdsource tasks on their behalf. The motivation for this new broker based model can be manifold. As an example, brokers allow large businesses and corporations to crowdsourcing tasks without having to worry about framing and posting tasks to crowdsourcing marketplaces.

The transformation of how people collaborate and interact on the Web has been poorly leveraged in existing service-oriented architectures. In SOA, compositions are based on Web services following the loose coupling and dynamic discovery paradigm. In this work, we highlighted the role of humans in SOA as *first class citizens*. We argue that people should be able to define interaction interfaces (services) following the same principles to avoid the need for parallel systems of Software-Based Services (SBS) and Human-Provided Services (HPS). We define such systems as mixed service-oriented systems.

The benefit of this approach is a seamless service-oriented infrastructure of human- and software-based services. In this research, we focus on innovative applications based on mixed service-oriented systems. Specifically, we focus on service-oriented crowdsourcing in open Web-based environments. The most prominent crowdsourcing platform is currently Amazon Mechanical Turk. An application of crowdsourcing is to outsource tasks that are difficult to implement as solutions based on software services. Another benefit of crowdsourcing is the on-demand allocation of a flexible workforce. Dynamically changing properties including user preferences,

D. Schall, *Service-Oriented Crowdsourcing*, SpringerBriefs in Computer Science, DOI: 10.1007/978-1-4614-5956-9_5, © The Author(s) 2012

changing expertise, and reputation make the design of mixed service-oriented systems challenging. The novelty of our approach is that context-sensitive interaction mining algorithms track these properties based on monitoring of ad-hoc interactions.

Finally, human-interactions are a substantial part of today's business processes. It becomes increasingly important to enable human-interactions in service-oriented systems. This has led to specifications such as WS-HumanTask and BPEL4People which aim at standardizing the interaction protocol between software processes and humans. These specifications received considerable attention from major industry players due to their extensibility and interoperability. Most efforts to model human interactions using BPEL4People focus on relatively static role models for selecting the right person to interact with. Thus, BPEL4People is not well suited for specifying and executing processes involving crowdsourcing of tasks to online communities. Here, we extended BPEL4People with non-functional properties that allow to cope with the inherent dynamics of crowdsourcing processes. Such properties include human capabilities and the level of skills. We discussed the formation of social networks that are particularly beneficial for processing extended BPEL4People tasks.